'99

in print paper only

12.95

mark on lower edge.

BEGINNING AGAIN

BEGINNING AGAIN

People and Nature
in the New Millennium

DAVID EHRENFELD

NEW YORK   OXFORD
Oxford University Press
1993

Oxford University Press

Oxford   New York   Toronto
Delhi   Bombay   Calcutta   Madras   Karachi
Kuala Lumpur   Singapore   Hong Kong   Tokyo
Nairobi   Dar es Salaam   Cape Town
Melbourne   Auckland   Madrid

and assoicated companies in
Berlin   Ibadan

Copyright © 1993 by Oxford University Press, Inc.

Published by Oxford University Press, Inc.,
200 Madison Avenue, New York, New York 10016

Oxford is a registered trademark of Oxford University Press

Library of Congress Cataloging-in-Publication Data
Ehrenfeld, David.
Beginning again : people and nature in the new millennium /
David Ehrenfeld
p.   cm.   Includes bibliographical references and index.
ISBN 0-19-507812-8
1. Human ecology—Philosophy.   2. Environmental protection.
3. Environmental policy.   I. Title.
GF21.E47   1993   304.2'8—dc20   92-14538

Many of the chapters in this volume are much revised from the originally published versions, and some titles have been changed. We gratefully acknowledge permission to reprint them.

"The Overmanaged Society" first appeared as "Environmental Control and the Decline of Reality" in *The Modern Churchman*, Vol. XXXII (2), 1990. "The Roots of Prophecy: Orwell and Nature" and "The Lesson of the Tower" first appeared in *The Hudson Review*, Volumes XXXVIII, 1985, and XXXIX, 1986, respectively. "Ecosystem Health" was reprinted by permission from *Ecosystem Health: New Goals for Environmental Management*, edited by R. Costanza, B. Norton, and B. Haskell. Copyright © 1992. Published by Island Press, Washington, D.C., and Covelo, California. "Hard Times for Diversity" is a rewritten version of two previously published essays, "Hard Times for Diversity," which first appeared in *Conservation for the Twenty-first Century*, by David Western and Mary Pearl, Copyright © 1989 by David Western and Mary Pearl, and reprinted by permission of Oxford University Press, Inc.; and "Why Put a Value on Biodiversity?" reprinted with permission from *Biodiversity*, © 1988 by the National Academy of Sciences. Published by National Academy Press, Washington, D.C. "Down From the Pedestal—A New Role for Experts" first appeared as "Environmental Protection: The Expert's Dilemma: in *Report from the Institute for Philosophy and Public Policy*, Vol. 11(2), 1991. It is based on a lecture given to the Oregon Chapter of the American Fisheries Society, in 1991. "Life in the New Millennium" first appeared as "Life in the Next Millennium: Who Will Be Left in the Earth's Community?" in *The Last Extinction*, edited by L. Kaufman and L. Mallory. Copyright © 1986. Reprinted by permission of MIT Press. "The Technology of Destruction" first appeared as "The Science of Edward Teller" in the August, 1985, issue of *California* magazine. "Asking Unaskable Questions—A Case Study" first appeared as "Garrett Hardin Is the Original Thinker's Original Thinker" in the November, 1984, issue of *California* magazine. "Changing the Way We Farm" is a rewritten version of two previously published essays, "Sustainable Agriculture and the Challenge of Place," which first appeared in *American Journal of Alternative Agriculture*, Vol. II(4), 1987; and "Changing the Way We Farm," which first appeared in *Orion* magazine, Spring, 1990. The remaining chapters, "Places," "Forgetting," "Dent de Lion—The Lion's Tooth," "After Valdez," "The Rain Forest of Selborne," "Desert Life," "Rights," "Loyalty," "State of the Art," and "A Turtle Named Mack," were first published by The Myrin Institute in *Orion* magazine, 1989-1992. Special thanks for their cooperation and generosity in letting me reprint all of them. In addition, excerpts from Dr. Seuss's books were reprinted by permission of Random House, Inc. and the Estate of Dr. Seuss. (*Yertle the Turtle and Other Stories* by Dr. Seuss. Copyright © 1950, 1951, 1958 by Theodor S. Geisel and Audrey S. Geisel. Copyright renewed 1977, 1979, 1986, by Theodor S. Geisel and Audrey S. Geisel. *Bartholomew and the Oobleck* by Dr. Seuss. Copyright © 1949 by Dr. Seuss. Copyright renewed 1976 by Theodor S. Geisel and Audrey S. Geisel.)

2 4 6 8 9 7 5 3 1

Printed in the United States of America
on acid-free, recycled paper

for Kate, Jane, Jonathan,
and Samuel

I call heaven and earth to witness against you this day: I have put before you life and death, blessing and curse. Therefore choose life, if you and your offspring would live.

Deuteronomy 30:19

Preface

Elijah has always been the most popular of the Hebrew prophets, largely because of his moral vision and effective, daring leadership during a time of overwhelming evil and social collapse, brought on by the amoral tyranny of King Ahab. When the prophet had completed his mission and was ready to bring his career on earth to an end, he did it in a characteristically theatrical way, by ascending to heaven in a chariot of fire drawn by horses of fire, all in the midst of a whirlwind. Nobody else in biblical history ever made this kind of exit, and Elijah's disciple and successor, Elisha, who witnessed the event, appears to have been apprehensive about following in such enormous footsteps. But Elijah was considerate of Elisha's feelings and left him his mantle, which had wondrous powers and which Elisha promptly used to work a miracle and demonstrate his succession as prophet.

The twentieth century is drawing to a close amid signs and portents of great impending change. Human population, powered by an unforgiving, ill-adapted, and poorly functioning technology, is rapidly growing past the inevitable crash point. It strains all resources of a tiring planet, as forests, fresh water, and agricultural soils are consumed and lost. In the process, other species and their habitats are destroyed, the very atmosphere and climate are altered, and the cultural and ethical heritage of humanity is replaced by barbarism and moral anarchy. Meanwhile, many of those who still have material wealth deny all change and use their powe. .o resist the efforts of that growing but still small number of people who are trying to find a better and more lasting way to live. In these turbulent, chaotic days, self-proclaimed prophets abound, some crying doom, some offering quick spiritual redemption, some announcing that the golden age of technology and human power is just around the corner. Now, when we need it most, the vision and hope of Elijah, and his mastery of terrible situations, seem to be absent, nor does he appear to have left any more mantles behind for us to take and put on.

False assumption and pretense are the enemies of prophecy, and our society is increasingly built of these destructive elements. The tragic discovery made by King Lear was that in a corrupted society, vision and understanding can only come to those who are somehow distanced from the corruption—distanced by wilderness, distanced by blindness, distanced by madness. Lear would not have found his vision today. The twentieth century coopts its visionaries. The motorcycles and dune buggies would chase Lear out of the wilderness; the solitude of his blindness would be invaded by blasting radios; his madness would be treated with antidepressants or, more likely, he would be victimized in dangerous shelters for the homeless.

Is it possible to steer a safe course through all this upheaval and chaos? Can we keep our distance from the rocks and reefs that lie all about us? I believe so: the secret, as in all navigation, is in paying attention to the fixed landmarks, both celestial and earthly. First are the spiritual landmarks—nearly all religious and ethical systems have them, to guide relationships among people and the relationship between people and a higher power. Spiritual landmarks are essential for our survival, but they are not enough.

The great rediscovery of twentieth-century ecology is that nature offers us a second, equally necessary, and infallibly useful kind of landmark: the innumerable examples of how to live and endure in the kaleidoscopic environment of our earthly and only home. To the imitation of God has been added the imitation of nature.

The two kinds of imitation have important similarities. Both should be learned starting at an early age; they can be learned later in life but the process is more difficult and less satisfying. And both kinds of imitation involve a sacrifice of immediate gratification; there are no appealing, easy, comfortable shortcuts to the ultimate felicity that the imitation of God and the imitation of nature provide. Moreover, the similarities are not coincidental: the spiritual landmarks and the natural landmarks are part of one grand, indissoluble system of guidance, the only system of guidance that works. This is not a new understanding. One example: an ancient midrash, or moral commentary on the Bible, found in the Jerusalem Talmud, states that Moses and David first had to be shepherds before they were fit to be leaders of Israel.

The Age of Hubris in which we now reside has obscured both kinds of landmark, especially the natural ones. At least we still pay lip service to the spiritual landmarks. Nature, on the other hand, exists to be conquered or ignored, not imitated. Those who still do learn from nature find it a slow business, full of surprises. In the pages that follow, for instance in the chapter on farming, I cite some practical examples. Here is one that is more whimsical but no less significant. In his book *Jungle Days,* first published in 1923, the naturalist William Beebe wrote a fascinating and amusing chapter on sloths (their near-total insensitivity to strong stimuli, their incredibly lethargic courtship, the minimalist bonding of sloth mother to child) and concluded with these words:

> It is interesting to know of the lives of such beings as this—chronic pacificists, normal morons, the superlative of negative natures, yet holding their own amidst the struggle for existence. Nothing else desires to feed on such coarse fodder, no other creature disputes with it the domain of the under side of branches. . . . From our human point of view sloths are degenerate; from another angle they are among the most exquisitely adapted of living beings. If we humans, together with our brains, fitted as well into the possibilities of our own lives we should be infinitely finer and happier. . . .

How to fit into the possibilities of our own lives is what the world of nature has to teach us, if we still have the wit and character to learn. For the people of the next millennium, the qualities of nature—honesty, reliability, durability, beauty, even humor—will be necessary landmarks for survival, there for the finding, unless the damage we are doing now proves too great. Yet that amount of damage need not occur. If we jettison our illusions and false assumptions of power and control soon, we *will* leave intact enough of the natural world to care for those who follow, as the ravens once cared for Elijah in the wilderness. Only then will we have a chance of discovering that Elijah left behind another mantle, after all.

I have been aided in this writing of this book by many people, and I apologize for not mentioning all of them. Among the friends who have influenced me the most are Wendell Berry, Wes Jackson, and David Orr, writers and teachers of unusual power and insight. Their humanity, wisdom, and moral vision have inspired me to my best efforts.

The late Archie Carr, now greatly missed, was my first teacher of conservation. When I read over what I have written, I see the many places in which his ideas appear: I hope I have done them justice.

One of my greatest joys in being a parent is to be helped in my work by my children. My oldest child, Kate, gave me the idea for the first chapter, and criticized and repaired the writing and sense of many others. I can count on her for supportive and good advice. My daughter Jane, a baby when I wrote *The Arrogance of Humanism,* is now old enough to give me ideas that I have incorporated with pleasure in these pages. Jon and Sam already give promise of future contributions.

The students in my conservation ecology classes have always provided a steady source of information and commentary to enhance what I think are my own thoughts but which at times must really be theirs.

More than half of the following chapters first appeared in the pages of *Orion* magazine, where they were greatly improved by the informed, sympathetic, and skilled editing of the managing editor, Aina Niemela. I was particularly fortunate that she was also willing to serve as general copy editor for the entire book, helping immeasurably to make it a more cohesive work. My editor at Oxford, Kirk Jensen, has been enthusiastic and encouraging from the beginning. And Sue Wigderson, my assistant at the journal I edit, *Conservation Biology,* has cheerfully and efficiently relieved me of many tasks that would have slowed my writing if I had had to do them myself.

The greatest help has come, as always, from my wife, Joan. There is hardly a word I have written or an idea I have expressed that has not had the benefit of her extraordinary critical analysis. Her patience with me is almost as remarkable as her ability to find time for this labor of love in spite of the twin pressures of her own demanding work and a demanding family. Of course, neither she nor the others I have thanked are responsible for any errors that may remain.

Highland Park, New Jersey   D. E.
May 1992

Contents

I TAKING BEARINGS

Places	3
The Roots of Prophecy: Orwell and Nature	8
The Rain Forest of Selborne	29
Desert Life	34
A Turtle Named Mack	41

II OFF COURSE

The Overmanaged Society	49
Forgetting	65
State of the Art	73
The Lesson of the Tower	79
After Valdez	95
Dent de Lion—The Lion's Tooth	99
The Technology of Destruction	104
Hard Times for Diversity	114

III TRUE HEADING

Rights	127
Loyalty	132

Ecosystem Health 139
Down from the Pedestal—A New Role for Experts 147
Asking Unaskable Questions—A Case Study 158
Changing the Way We Farm 163
Life in the New Millennium 175

Suggested Readings 195
Index 205

I

Taking Bearings

God did not make this world in jest; no, nor in indifference. These
migrating sparrows all bear messages that concern my life.

HENRY DAVID THOREAU
Journal, Vol. III, March 31, 1852

Places

An awareness of places doesn't always come naturally. In my case it was definitely an acquired trait. The place was Tortuguero, in Costa Rica. I had gone there as a graduate student, sent by my major professor, the late Archie Carr. He was one of the great conservationists of our time and a remarkable teacher, but he didn't teach conservation by any formal process of instruction. He was too smart for that. Instead, he taught about places through direct experience and let his students' feelings of conservation develop spontaneously, as they almost invariably did.

The beginning of my introduction to places was not in Tortuguero at all, but in Gainesville, Florida, where I had come to study zoology with Dr. Carr. It was my first day there, and it was summer, hot and humid. My new medical diploma from Harvard was in my suitcase and I was wearing a tie and jacket. I was terribly out of place, the way only a person who doesn't have a good feeling for places can be.

Archie greeted me as if he had been waiting eagerly for my arrival for years. (Even after I had been around long enough to learn that he greeted everybody that way, those greetings still made me feel good.) When the formalities were over, a matter of maybe a minute, he said:

"Ehrenfeld, have you ever seen an alligator nest?"

I hadn't. There weren't any in Boston. Nor in northern New Jersey, where I grew up.

In five minutes we were plunging through the vegetation along the shore of Lake Alice, at the southern end of the University of Forida's campus. I was still wearing my jacket and tie. The alligator nest, when we reached it, looked like a haphazard pile of mud and sticks. I don't remember whether Archie uncovered any of the elongated white eggs to show me; probably he did. What I do remember is that he pointed out the mother alligator floating about fifty feet offshore, and told me to watch her. Then, while I watched, he began to make a soft, croaking, chuckling, grunting noise.

Soon, the alligator swung around toward us and moved closer to shore.

"That's the noise alligator hatchlings make," said Archie. "When they hatch out, the mother hears them and comes to release them from the nest and protect them. They stay with her a year or two. Old E. A. McIlhenny, the Louisiana tabasco sauce king, was the first person to write about the fierce maternal behavior of alligators, back in 1935. He wasn't a trained zoologist. Nobody believed him. Reptiles weren't supposed to be maternal. But everything he said was right."

He resumed croaking. The alligator came a little closer and then stopped. Apparently she was getting used to Professor Carr's impersonations, and was no longer impressed. After a while, we left.

That night, as I was dropping off to sleep, I wondered why Archie had taken me to look at an alligator nest before introducing me to his colleagues or showing me around the lab. I couldn't figure it out. Only years later did it occur to me that this was a proper introduction to a place—visiting a typical part of the landscape and meeting its oldest inhabitants. Alligators, after all, are just an advanced kind of dinosaur. And knowing Archie and his delight in the comic, he probably also wanted to see me in my tie and jacket up against that alligator nest. But at the time I wasn't aware of these things. It took another kind of reptile to teach me about places and conservation.

It was Archie's intention, and mine, that I would do my doctoral research on the orientation and navigation of sea turtles, the so-called green turtle (*Chelonia mydas*), at Tortuguero, the great nesting beach along Costa Rica's Caribbean coast. Archie's research had proved that green turtles are philopatric nesters; that is, they return to the same place to nest every two, three, or four years, often swimming many hundreds of miles to get there. Any turtle we found nesting at Tortuguero we measured, described, and marked with a numbered metal tag attached to a front flipper and bearing Archie's address and an offer of a reward for its return. That way we recognized turtles who had been on the beach before, and the tag returns from turtle fishermen told us where the turtles went after they left Tortuguero.

Because the turtles come out to nest after dark, much of my work was done at night. There was a great deal of waiting between turtles, plenty of time to sit on a driftwood log and think. In the first years of my research I was often the only one on the beach for miles. After ten or twenty minutes of sitting without using my flashlight, my eyes adapted to the dark and I could make out forms against the brown-black sand: the beach plum and coconut palm silhouettes in back, the flicker of the surf in front, sometimes even the shadowy outline of a trailing railroad vine or the scurry of a ghost crab at my feet. The air was heavy and damp with a distinctive primal smell that I can remember but not describe. The rhythmic roar of the surf a few feet away never ceased—my favorite sound. I hear it as I write in my landlocked office in New Jersey. And then, with ponderous, dramatic slowness, a giant turtle would emerge from the sea.

Usually I would see the track first, a vivid black line standing out against the lesser blackness, like the swath of a bulldozer. If I was closer, I could hear the animal's deep hiss of breath and the sounds of her undershell scraping over logs. If there was a moon, I might see the light glistening off the parabolic curve of the still wet shell. Size at night is hard to determine: even the sprightly 180-pounders, probably nesting for the first time, looked big when nearby, but the 400-pound ancients, with shells nearly four feet long, were colossal in the darkness. Then when the excavations of the body pit and egg cavity were done, if I slowly parted the hind flippers of the now-oblivious turtle, I could watch the perfect white spheres falling and falling into the flask-shaped pit scooped into the soft sand.

Falling as they have fallen for a hundred million years, with the same slow cadence, always shielded from the rain or stars by the same massive bulk with the beaked head and the same large, myopic eyes rimmed with crusts of sand washed out by tears. Minutes and hours, days and months dissolve into eons. I am on an Oligocene beach, an Eocene beach, a Cretaceous beach—the scene is the same.

It is night, the turtles are coming back, always back; I hear a deep hiss of breath and catch a glint of wet shell as the continents slide and crash, the oceans form and grow. The turtles were coming here before here was here. At Tortuguero I learned the meaning of place, and began to understand how it is bound up with time.

I also learned what it means to have to stand by while a place is damaged. What it means to watch while the forest behind the beach is exploited by ignorant opportunists with foreign capital. What it means to watch while a tagged turtle, one we have found nesting at the same spot three times in a decade, crawls slowly down the beach back to the water, where a dozen turtle fishing boats are waiting just beyond the surf.

Although my awareness of the need for conservation came slowly as a result of many experiences, I associate it especially with one night at Tortuguero. There were four or five of us walking the beach, including Archie and José Figueres, better known as "Don Pepe," the leader of the revolution, liberator of Costa Rica, and first president of the only nonmilitarized democracy of any size on the planet. There was no guard; none was needed.

It was Don Pepe's first visit to the legendary Tortuguero—we had been watching a green turtle nest, also a first for him. El Presidente, a short, Napoleonic man with boundless energy, was enjoying himself enormously. Both he and Archie were truly charismatic people, and they liked and respected each other. Each was fluent in the other's language. The rest of us went along quietly, enjoying the show. As we walked up the beach toward the boca, where the Rio Tortuguero meets the sea, Don Pepe questioned Dr. Carr about the green turtles and their need for conservation. How important was it to make Tortuguero a sanctuary? Just then, a flashlight picked out a strange sight up ahead.

A turtle was on the beach, near the waterline, trailing something. And behind her was a line of eggs which, for some reason, she was depositing on the bare, unprotected sand. We hurried to see what the problem was.

When we got close, it was all too apparent. The entire undershell of the turtle had been cut away by poachers who were after the calipee, or cartilage, to dry and sell to the European turtle soup manufacturers. Not interested in the meat or eggs, they had evidently then flipped her back on her belly for sport, to see where she would crawl. What she was trailing was her intestines. The poachers had probably been frightened away by our lights only minutes before.

Dr. Carr, who knew sea turtles better than any human being on earth and who had devoted much of his life to their protection, said nothing. He looked at Don Pepe, and so did I. It was a moment of revelation. Don Pepe was very, very angry, trembling with rage. This was his country, his place. He had risked his life for it fighting in the Cerro de la Muerte. The turtles were part of this place, even part of its name, Tortuguero; they had been coming here long before people existed in Central America. He understood that, just as he understood the profound significance of the useless, round, white eggs swept by the retreating wavelets down the packed sand into the surf beyond. No green turtle born at Tortuguero will ever lay her eggs anywhere else. She was home, laying her eggs for the last time.

Today there are far more green turtles at Tortuguero than there were when Don Pepe made his first visit, thanks to him, to other enlightened citizens of his remarkable country, and above all to Archie Carr. I haven't been back since 1983. No time, the cost of flying a family of six to San José—these are the sorts of excuses I make. But the real reason is more selfish. How will Tortuguero be with electricity and vacation cabins, with a road to Tortuguero Village, without Archie or Sheftan Martinez or Bertie Downs or Miss Sibella, and with all those eager new conservationists crowding the black beach at night? Will it be the same place? Will I enjoy being there?

Of course I know the answer is yes. Places can be destroyed, that is, they can have their nature and meaning irrevocably changed and their connection with the past severed. All conservationists are aware of that. This hasn't happened at Tortuguero. It has changed, but it is still the same place. When I do go back, as I must because places get in your blood, I know I will still be able to find a log on the beach to sit on alone in the darkness, and with luck will see a dim, rounded form heaving itself out of the nearby sea.

The Roots of Prophecy: Orwell and Nature

Futurology notwithstanding, nobody can foretell the future. That does not mean that there are no true prophets—only that our understanding of what is prophecy has become confused. In modern times we have had our share of prophets. Among those who have written in English, names such as C. S. Lewis, E. F. Schumacher, James Burnham, and Lewis Mumford would be prominent on lists of prophets, although not always together on the same lists. But one name would be on nearly every list, the name of George Orwell. As we try to cope with this most complex, unstable, and unpredictable era of human history, it may help to learn how a master navigator took his bearings.

Oddly enough, most people who associate Orwell with the idea of prophecy do so because of *1984* and *Animal Farm,* while the main body of his prophetic writings, the essays and the remaining novels, are known to a much smaller group of readers. To the majority, then, who search *1984* for a description of the present, Orwell is bound to disappoint as prophet; many of the special features of *1984* remain fictional. Even in the erstwhile Soviet Union at the peak of its power, the houses were not equipped with two-way video sets monitored by the KGB, and no technology of brainwashing was able to overcome the spirit of a Mendelevich, a Scharansky, or a Sakharov.

Moreover, unlike the script of *1984*, world power has not been partitioned among three superstates. Instead we have seen the giant Russian empire disintegrating into a chaotic collection of alternately repelling and attracting fragments, the United States struggling to preserve the dwindling remnants of its much-abused wealth and spirit, and China wallowing in a morass of people and environmental devastation. A mere handful of terrorists and kidnappers can throw the most powerful nations on earth into paroxysms of impotent rage. In the largest countries, state and private purveyors sell sophisticated weapons to their enemies even as they fight them. None of this is in *1984*.

Nevertheless, Orwell was a prophet in the classical sense. The biblical prophets themselves were seldom the kind of holy for-tunetellers that the word *prophet* has come to signify. According to Exodus, Moses knew that God could be trusted to care for the children of Israel in the wilderness, yet he did not have the faintest inkling beforehand that they would be fed with quails and manna.

The business of prophecy is not foretelling the future; rather it is *describing the present* with exceptional truthfulness and accuracy. Once this is done—and it is an overwhelmingly difficult task—then it can be seen that certain broad aspects of the future have become self-evident, while other features, including many of the details, remain shrouded in mystery. This is prophecy.

Describing the present so that it reveals part of the future is an act that requires far more than intellect in all but the simplest instances. A mere catalog of facts is not a useful description of the present (or any good computer could be programmed for prophecy); the facts must be sorted subjectively and selected by a human being who is skilled at such things. But this subjective sorting, although neces-sary, is fraught with danger. In most of us, the process is confounded by emotions or feelings—love, hatred, fear, envy—of which we may be only dimly aware. What emerges is not a description of the present but the portrait of a personal fantasy of the present.

Orwell never rejected the emotional parts of thought and analysis. Indeed to do so, assuming that it could be done, would render the products of such thought worthless and inane. Rather, he said that for analysis to succeed in mirroring and clarifying reality, the emotional, subjective side of it cannot be buried in the unconscious mind of the analyzer. In his "London Letter" to *Partisan Review,* written in the winter of 1944, Orwell wrote with his characteristic understatement:

> I believe that it is possible to be more objective than most of us are, but that it involves a *moral* effort. One cannot get away from one's own subjective feelings, but at least one can know what they are and make allowance for them.

To give a brief example, it is impossible to read Orwell's writings without perceiving a faint but evident strain of anti-Semitism. Yet, remarkably, in 1945 he took the time to gather up the threads of his own anti-Semitism, examined them critically as if they belonged to someone else, and used them as material for one of the finest essays ever written on the subject. It is easy to be objective about such topics as sea urchins or the square root of two, Orwell noted, but not about things that personally matter.

> What vitiates nearly all that is written about anti-Semitism is the assumption in the writer's mind that *he himself* is immune to it. "Since I know that anti-Semitism is irrational," he argues, "it follows that I do not share it." He thus fails to start his investigation in the one place where he could get hold of some reliable evidence—that is, in his own mind.

There is another essential ingredient of prophecy besides brains and the capacity to examine critically one's own subjective feelings: it is a sense of responsibility for the words one uses. Orwell had that, too. In 1941, in his essay called "The Lion and the Unicorn," he criticized his fellow socialists and left-wing intellectuals as having "little in them except the irresponsible carping of people who have never been and never expect to be in a position of power." Until the publication of the American edition of *Animal Farm,* a few years before his death, Orwell himself had no power of any consequence—yet he weighed and measured his words with a sense of responsibility that could have been no greater if he had been prime minister.

But if these are the necessary preconditions of prophecy, they are by themselves not sufficient to ensure its greatness. For that we have to look beyond the person for an independent standard against which things and events can be measured. Just as the internal biological clock of an animal or plant must be "set" by some independent time-giving event such as sunrise, so the prophet needs a proper set of weights and measures against which to balance and compare the happenings of the world. For the biblical prophets, this independent standard was the word of God as embodied in the commandments. Some of the modern prophets—C. S. Lewis and E. F. Schumacher, for example—have used the same standard. George Orwell, who did not believe in God, did not take this option. Yet the brilliance of his prophecy is evident: one only has to recognize the sordid Orwellian newspeak that has been an integral and necessary part of the weapons race, the promotion of consumerism, the genocide of native peoples, the selling of nuclear power, the centralization of government, the proliferation of experts, and the debasing of human environments—all of which were implicitly or explicitly described in Orwell's writings—to see his prophetic genius.

If not the word of God, what was the external standard that enabled Orwell to describe his world and ours with such faithfulness to present and future reality? For anyone who ventures beyond *1984* in Orwell's writings, this is not a hard question to answer. Fortunately for us, for we need prophets in this age of overwhelming and turbulent change, Orwell had a profound understanding both of nature and of humanity's place in it. He knew nature from encounter, not from intellect. Awareness of the natural world was with him constantly. Nature was his independent standard, the open secret of his prophetic vision.

What was nature to George Orwell? Simply put, I believe it was his model, his example of certain qualities that he wanted to emulate and did emulate in his life and in the art that was his work. They are also the qualities that he used as a kind of standard, a measuring stick, in his analysis of the changing life of his and our times. There are at least three of these qualities of nature that seem to have been important to Orwell—my own arbitrary categories, it's true, but perhaps no less real for that. The first of these qualities or properties of nature is *honesty;* the second I have to describe with four different but related words, *reliability/continuity/durability/resilience;* the third property of nature that was important to Orwell is its *beauty* and *serenity.*

Honesty

Honesty was the hallmark of Orwell's personal style and method. In his essay called "Why I Write," he noted that from the beginning of his career he had "a power of facing unpleasant facts." Often these facts were about himself, his own behavior. We see this in his autobiographical writings, such as the following brief passage from *Down and Out in Paris and London*. Orwell was working as a dishwasher in a horrible Parisian restaurant; the hours were very long, the pay was minuscule, everyone was always tired.

> We quarrelled over things of inconceivable pettiness. The dustbin, for instance, was an unending source of quarrels—whether it should be put where I wanted it, which was in the cook's way, or where she wanted it, which was between me and the sink. Once she nagged and nagged until at last, in pure spite, I lifted the dustbin up and put it out in the middle of the floor, where she was bound to trip over it.
>
> "Now, you cow," I said, "move it yourself."
>
> Poor old woman, it was too heavy for her to lift, and she sat down, put her head on the table and burst out crying. And I jeered at her.

How many of us living through such a scene would remember it that way, let alone put it down in indelible print for anyone to read? Nature, Orwell's external standard of honesty, also never varnishes itself: it is exactly what it is, always. One doesn't look at a beech woods, a marsh, or a stream and think, "It's lying to me." We encounter an awareness of nature's honesty everywhere in Orwell's writing, but nowhere more vividly expressed than in his wonderful, imperfect novel, *Coming Up For Air*. In this novel the hero, whose name is also George, describes his love of fishing as a child, and the thrill of his discovery of a tiny deep pool that nobody else knew about, one that contained enormous fish.

> You had to fight your way through a sort of jungle of blackberry bushes and rotten boughs that had fallen off the trees. I struggled through it for about fifty yards, and then suddenly there was a clearing and I came to another pool which I had never known existed. . . . It was very clear water and immensely deep. . . .

I hung about for a bit, enjoying the dampness and the rotten boggy smell. . . . And then I saw something that almost made me jump out of my skin.

What he saw was a school of gigantic carp.

As things turn out, he never gets a chance to fish for them until as a middle-aged man he returns home after an absence of thirty years and visits the old estate whose grounds the pool was on. The manor house is now an insane asylum, and the grounds have become a separate suburban housing development. He meets a friendly old man who tells him:

> "We live in the midst of Nature up here. No connection with the town down there"—he waved a hand in the direction of Lower Binfield—"the dark Satanic mills—te hee!" . . .
> Immediately, as though I'd asked him, he began telling me all about the upper Binfield Estate, and young Edward Watkin, the architect, who had such a feeling for the Tudor, and was such a wonderful fellow at finding genuine Elizabethan beams in old farmhouses and buying them at ridiculous prices.

Of course the pool is now a garbage dump, drained and already half-filled with tin cans. But the residents have preserved a small patch of the original woods. The old man says:

> "That is sacrosanct. We have decided never to build in it. It is sacred to the young people. Nature, you know. . . . We call it the Pixy Glen."
> The Pixy Glen. . . . And they'd filled my pool up with tin cans. God rot them and bust them! Say what you like—call it silly, childish, anything—but doesn't it make you puke sometimes to see what they're doing to England, with their birdbaths and their plaster gnomes, and their pixies and tin cans, where the beechwoods used to be?

Clearly what bothered Orwell here, even more than the pollution and the rubbish and the destruction, was the dishonesty, the corruption of the inherent honesty of nature.

But nature to Orwell is not all beech woods and pristine pools—far from it. In the same book, George explains his uneasy feelings about the new, big, raw cemetery outside of town, away from the houses and stores.

> We had our churchyard plumb in the middle of the town, you passed it every day, you saw the spot where your grandfather was lying and where some day you were going to lie yourself. We didn't mind looking at the dead. In hot weather, I admit, we also had to smell them, because some of the family vaults weren't too well sealed.

The honesty of nature contrasted sharply with the new world that was beginning to emerge in the 1930s and which may not yet in our day have hit its full stride. A world in which two plus two may equal five, war is peace, and freedom slavery. It is no coincidence that in *1984,* the last honest free conversation held by Winston and Julia concerns nature.

> "Do you remember," he said, "the thrush that sang to us, that first day, at the edge of the wood?"
> "He wasn't singing to us," said Julia. "He was singing to please himself. Not even that. He was just singing."

Note that she takes great pains to be scrupulously honest about nature. Nature, like people, deserves nothing less than full honesty; it is itself the model of honesty. Seconds after this conversation takes place, the thought police burst into the room, and that for all practical purposes is the end of the story.

Illusion, unreality, deception, were becoming institutionalized, a part of everyday life. Orwell, using nature as his measuring stick, saw it very early, long before television had realized any of its potential as an instrument for the manipulation of truth. In the beginning of *Coming Up For Air* there is a scene in which George Bowling buys a frankfurter in a coffee shop and gets a surprise:

> The frankfurter had a rubber skin, of course, and my temporary teeth weren't much of a fit. I had to do a kind of sawing movement before I could get my teeth through the skin. And then suddenly—pop! The thing burst in my mouth like a rotten pear. A sort of horrible soft stuff was oozing all over my tongue. . . . It was *fish!* A sausage, a thing

calling itself a frankfurter, filled with fish! . . . That's the way we're going nowadays . . . everything made out of something else. Celluloid, rubber, chromium-steel everywhere, arc-lamps blazing all night, glass roofs over your head, radios all playing the same tune, no vegetation left, everything cemented over, mock-turtles grazing under the neutral fruit trees. But when you come down to brass tacks and get your teeth into something solid . . . that's what you get. Rotten fish in a rubber skin.

Reliability/Continuity/Durability/Resilience

I can't give more than a tiny sample of the many instances in which Orwell acknowledges and pays his respects to this side of nature. The seasons of nature—not a vague awareness but an incredibly acute perception—are always with him; indeed that awareness was one of his main anchors to windward.

From a letter to his friend Richard Rees, April 1936:

> It is still beastly cold and everything very late. I have found no nests except thrushes and blackbirds and have not heard the cuckoo or seen a swallow—I usually see my first about the 14th. The blackthorn is out and there are plenty of primroses and cowslips but the hedges are still very bare.

Part of the only surviving letter to his first wife, from a hospital bed. April 1937:

> The weather is much better, real spring most of the time, and the look of the earth makes me think of our garden at home and wonder whether the wallflowers are coming out.

Nothing can stop the cycle of nature. From an essay called "Some Thoughts on the Common Toad," 1946:

> As for spring, not even the narrow and gloomy streets round the Bank of England are quite able to exclude it. It comes seeping in everywhere, like one of those new poison gasses which pass through all filters.

Some of his most sparkling and cheerful writing was inspired by the events of the natural cycle. From the same essay on the common toad:

> Before the swallow, before the daffodil, and not much later than the snowdrop, the common toad salutes the coming of spring after his own fashion, which is to emerge from a hole in the ground, where he has lain buried since the previous autumn, and crawl as rapidly as possible towards the nearest suitable patch of water. . . . At this period, after his long fast, the toad has a very spiritual look, like a strict Anglo-Catholic towards the end of Lent.

In *Coming Up For Air,* George Bowling remembers the foods that grew wild in the hedges at different seasons:

> In July there were dewberries—but they're very rare—and the blackberries were getting red enough to eat. In September there were sloes and hazel-nuts. The best hazel-nuts were always out of reach. Later on there were beech-nuts and crab-apples.

It is important to point out that this is not wilderness that he is writing about. It is an English hedge, a form of agriculture, planted and maintained by skilled, hard-working people. Orwell was not a wilderness purist: he valued nature most when it had been shaped and molded by human beings who understood and respected it.

The durability of nature has a remarkable power to atone for, almost to blot out, human evil. Orwell describes this in his 1946 essay, "A Good Word for the Vicar of Bray." The Vicar of Bray, immortalized in a humorous folk song, was an eighteenth-century English churchman who became notorious for changing his religious and political allegiances to curry favor with every new government or king that came to power during his long tenure. Orwell visited his church and discovered that a magnificent yew tree, planted by the vicar, still survived.

> after this lapse of time, all that is left of him is a comic song and a beautiful tree, which has rested the eyes of generation after generation and must surely have outweighed any bad effects which he produced by his political quislingism. . . .

> The planting of a tree, especially one of the long-living hardwood trees, is a gift which you can make to posterity at almost no cost and with almost no trouble, and if the tree takes root it will far outlive the visible effect of any of your other actions, good or evil.

In that essay he describes his feelings upon returning to a cottage where he once lived, and looking at the roses, gooseberry bushes, and fruit trees he had planted ten years earlier. In particular, he was proud of one fruit tree, a Cox's Orange Pippin, the finest English variety of apple.

> I maintain that it was a public-spirited action to plant that Cox, for these trees do not fruit quickly and I did not expect to stay there long. I never had an apple off it myself, but it looks as if someone else will have quite a lot. By their fruits ye shall know them, and the Cox's Orange Pippin is a good fruit to be known by.
>   . . . A thing that I regret, and which I will try to remedy some time, is that I have never in my life planted a walnut. Nobody does plant them nowadays—when you see a walnut it is almost invariably an old tree. If you plant a walnut you are planting it for your grand-children, and who cares a damn for his grandchildren?

Orwell felt very strongly that there is something missing from modern life, and an understanding of nature told him what that something is. This is from *Coming Up For Air:*

> people then had something that we haven't got now . . . they didn't think of the future as something to be terrified of. It isn't that life was softer then than now. Actually it was harsher. People on the whole worked harder, lived less comfortably and died more painfully. . . . And yet what was it that people had in those days? A feeling of security, even when they weren't secure. More exactly, it was a feeling of continuity.

A few pages on he tells what replaced that continuity.

> There was a temporary feeling about everything.

And for life to produce such a feeling—Orwell doesn't have to put it in words—is unnatural, a departure from nature.

Beauty and Serenity

The third property of nature that Orwell valued was its beauty and
serenity. He was keenly attuned to it and found it in unlikely places.
Again from his essay on the common toad. When the toad emerges
from his winter burrow,

> his body is shrunken, and by contrast his eyes look abnormally large.
> This allows one to notice, what one might not at another time, that a
> toad has about the most beautiful eye of any living creature. It is like
> gold, or more exactly it is like the golden-coloured semi-precious
> stone which one sometimes sees in signet rings, and which I think is
> called a chrysoberyl.

Later in this essay, after dismissing the argument that only people
who live in cities and don't know its brutal side can love nature, he
states his idea of the usefulness of natural beauty.

> I have always suspected that if our economic and political problems
> are ever really solved, life will become simpler instead of more
> complex. . . . I think that by retaining one's childhood love of such
> things as trees, fishes, butterflies and—to return to my first in-
> stance—toads, one makes a peaceful and decent future a little more
> probable, and that by preaching the doctrine that nothing is to be
> admired except steel and concrete, one merely makes it a little surer
> that human beings will have no outlet for their surplus energy except
> in hatred and leader worship.

I want to stress once more that Orwell is not talking about wilderness so much as environments in which people and nature interact in
a positive way. Ecologists who study natural communities have only
begun to appreciate the characteristics and richness of this kind of
environment.

Natural beauty for Orwell is a subjective quality perceived by
people in places where they live and work, and doesn't only pertain
to something you might want to photograph for a picture postcard. In
his odd and revealing novel *A Clergyman's Daughter,* Orwell has a
memorable chapter in which the daughter, a victim of amnesia, has

wandered off to pick hops with a motley assortment of gypsies, tramps, migrant farmworkers, and Cockney tradespeople down from London. He wrote the scene from personal experience described elsewhere in his notes and diaries. The pickers, who slept in piles of straw in drafty wooden shacks, got up at the first light of dawn, cooked their breakfast and dinner—bacon, bread fried in bacon grease, and tea, for both meals—ate the breakfast portion, and carried the dinner in a pail out to the hop fields a mile or two away. Moving the heavy bins into position, they dragged the equally heavy hop bines over them and

> began tearing off the heavy bunches of hops. At that hour of the morning you could only pick slowly and awkwardly. Your hands were still stiff and the coldness of the dew numbed them, and the hops were wet and slippery. . . .
>
> The stems of the bines were covered by minute thorns which within two or three days had torn the skin of your hands to pieces. In the morning it was a torment to begin picking when your fingers were almost too stiff to bend and bleeding in a dozen places; but the pain wore off when the cuts had reopened and the blood was flowing freely.

At noon they ate their bacon sandwiches and drank their tea, then picked until five or six in the evening. Wages were one and a half to two pence a bushel—you might earn about ten shillings a week, but some of this was likely to be withheld by the paymaster on one pretext or another. When it rained, which was often, you waited under the dripping bines unable to pick, and might get ten or eleven pennies for the day's work. Now keep all this in mind and read Orwell's description of the Cockneys who came from London in the hop-picking season:

> respectable East Enders, costermongers and small shopkeepers and the like, who came hop-picking for a holiday and were satisfied if they earned enough for their fare both ways and a bit of fun on Saturday nights.

Nowhere do I find any indication that this idea of a holiday surprised Orwell. He knew poverty well and had no rosy illusions about it. He lived in poverty for all but the last three and a half years of his life, and even then he was not by most standards a wealthy man. In 1948, for example, two years before his death, this greatest English essayist of his century wrote from a hospital tuberculosis ward to his friend Julian Symons to thank him for sending a ball-point pen to replace the one he had used up. Orwell never romanticized poverty: he merely understood that the seminatural greenery and quiet of a hop field on a chilly September morning was the only bit of beauty and peace that many people could ever have, and there is no hint in his account that he finds anything ludicrous, outrageous, or even sad in the idea of such a vacation.

In April of 1945, in a letter to his friend Anthony Powell, he described the sudden death of his first wife, Eileen, two weeks earlier, a casualty of anesthesia during a minor operation, and then mentions his adopted son, Richard, who was at that time about a year old.

> As soon as I can get a nurse and a house I shall remove him to the country, as I don't want him to learn to walk in London.

Why not, one might ask? After all, London is open for babies to walk in at all seasons. Surely he didn't mean that the child would learn to walk badly in London. I think the answer is plain: when his child made his first contact with the earth, without a mother, he wanted him to feel the beauty and serenity of a nurtured landscape.

Orwell's Contact with Nature

Before exploring further the implications of Orwell's feeling for nature, I want to make a point that might not have been obvious so far. Orwell's contact with nature was not primarily intellectual, and certainly not forced or contrived. Unlike Thoreau, whose odes to nature rang false to some of the great naturalists who knew his work, and whose genius for the appreciation and description of the natural world was occasionally marred by misinformation, Orwell was always in solid touch with the earth, even in London, and rarely made mistakes about what he saw. He, too, had an amazing facility for observation of animals and plants, but exercised this not while living alone in the woods or on trips through the wilderness; instead his experience of nature was gained almost entirely through gardening, or, more properly, small farming.

A letter to Eleanor Jaques, dated July 20, 1933, and signed Eric Blair, his real name:

> The heat here is fearful but it is good for my marrows and pumpkins, which are swelling almost visibly. We have had lashings of peas, beans just beginning, potatoes rather poor, owing to the drought, I suppose. I have finished my novel [*Burmese Days*].

A letter to Jack Common, from his new home at Wallington, April 1936:

> it isn't what you might call luxurious, but it is as good as one could expect for 7/6 a week so near London. The garden is potentially good but has been left in the most frightful state I have ever seen. I am afraid it will be a year before I can get it nice.

Orwell was in the Spanish Civil War and was wounded in the neck by a bullet and sent home. This is from a letter to Geoffrey Gorer, dated August 1937.

> The bullet went through my neck from front to back but skidded round both the carotid artery and the backbone in the most remarkable way. I have one vocal cord paralysed. . . . I am getting pretty well going with my book [*Homage to Catalonia*] and we are very busy trying to do something about the garden, which was in a ghastly mess when we got back.

Lest one think that Orwell planted only vegetables, the following is an extract from another letter to Jack Common, from French Morocco, where Orwell was recovering from one of his bouts with tuberculosis in December 1938.

> The effects of the frost were very curious. Some nasturtiums I had sown earlier were withered up by it, but the . . . Bougainvillea, which is a tropical plant from the South Pacific, weren't affected.

Bougainvillea is in fact South American, but perhaps Orwell can be forgiven this rare slip.

No conditions of poor health (save for outright confinement in a hospital bed) or societal upheaval could keep Orwell from planting and cultivating the earth. These activities were conditions of his existence, like breathing; there was nothing optional about them. From the wartime diary, April 22, 1941, during the bombing of Britain:

> Have been 2 or 3 days at Wallington. Saturday night's blitz could easily be heard there. . . .
> Sowed while at Wallington 40 or 50 lb of potatoes, which might give 200 to 600 lb according to the season, etc. It would be queer . . . if, when this autumn comes, those potatoes seem a more important achievement than all the articles, broadcasts, etc I shall have done this year.

From the only surviving letter to his second wife, Sonia Brownell, dated April 1947, two years before their marriage, from the Isle of Jura, where he had moved the day before, after being sick in bed:

> hardly a bud showing and I saw quite a lot of snow yesterday. However it's beautiful spring weather now and the plants I put in at the new year seem to be mostly alive. . . . I'm still wrestling with more or less virgin meadow, but I think by next year I'll have quite a nice garden here.

And finally, another letter from Jura, to Julian Symons, October 1948:

> Richard is blooming. He is still I think backward about talking, but lively enough in other ways and really almost helpful about the farm and garden. Something tells me he won't be one for book-learning and that his bent is for mechanics. I shan't try to influence him, but if he grew up with the ambition of being a farmer I should be pleased. Of course that may be the only job left after the atom bombs.

Richard Blair, whose father, George Orwell, died when Richard was about six years old, became a farmer. As for Orwell, the only thing that ever effectively kept him from working the land and from observing and loving nature was death itself.

God and Nature

Having seen the depth of Orwell's commitment to nature, the logical question one might ask is whether Orwell's feelings about the natural world had a religious side, whether nature was for him in any way an expression of or substitute for the Deity. At the outset I should say that there is not a shred of evidence to indicate that this is the case. To anyone who reads Orwell with an effort at objectivity, it should be clear that at no time during his adult years did he worship God in nature, nor did he worship nature instead of God, nor did he worship God or nature in any other way, even a personal, idiosyncratic way. He didn't worship. Not God, not nature, not power, not money, not beauty, not love, not even art. Not anything. He was not a worshipping type.

Some see his failure to believe in God as the principal flaw in his character; however it would be difficult to show that Orwell's lack of belief marred his writing or, for that matter, his gift of prophecy. Indeed his lack of belief may actually have helped increase his impartiality in the emotionally charged arena of formal religion. Moreover, no one can accuse Orwell of insensitivity to the meaning of God in the lives of other people. Here is what Dorothy, the clergyman's daughter, thinks as she struggles successfully to resolve her crisis of faith:

> Since you exist, God must have created you. . . . He created you, and He will kill you, for His own purpose. But that purpose is inscrutable. It is in the nature of things that you can never discover it, and perhaps even if you did discover it you would be averse to it.
> . . . There was, she saw clearly, no possible substitute for faith; no pagan acceptance of life as sufficient to itself, no pantheistic cheer-up stuff, no pseudo-religion of "progress" with visions of glittering Utopias and ant-heaps of steel and concrete. It is all or nothing. Either life on earth is a preparation for something greater and more lasting, or it is meaningless, dark and dreadful.

But Dorothy had lost her faith beyond all hope of recovery. Without faith, what alternative to black despair was there for her? The narrator describes yet a third alternative:

> She did not reflect, consciously, that the solution to her difficulty lay in accepting the fact that there was no solution; that if one gets on with the job that lies to hand, the ultimate purpose of the job fades into insignificance; that faith and no faith are very much the same provided that one is doing what is customary, useful and acceptable.

This was certainly Orwell speaking of and for himself. After he wrote that paragraph about getting on with the job, perhaps he went outside and planted a row of potatoes, a few hills of pumpkins, or even a Cox's Orange Pippin whose apples he would never get to taste.

Two Visions of Utopia

A life in which human beings live in some kind of uneasy, imperfect, yet generally wholesome, productive, and durable equilibrium with nature was for Orwell a vision of utopia, of paradise. But it wasn't his only vision of paradise. Unlike most of us who have such dreams, he had *two* visions, distinct although not unrelated, and these two views of a utopian future set up a critical tension in his writings, especially his social and political writings, which is the bulk of them. Orwell, a Socialist, also had a social vision of an ideal world in which no class of people or race or nationality would be exploited by any other, and in which people would share, more or less equally, whatever wealth might be forthcoming from a world that was not abused.

He knew every variety of exploitation from firsthand experience from a variety of perspectives: as a British colonial officer with the Imperial Police in Burma, as an anarchist soldier in the Spanish Civil War opposed to both Franco and the Communists, as a penniless tramp in France and England, and as an investigative reporter studying conditions in the British coal mines. He had a rare sympathy for the plight of others. But his honesty and detachment were always in evidence, regardless of his personal involvement. One would be hard pressed to find, for example, a more sensitive portrait of the degrading effects of colonialism than the story in *Burmese Days* of Dr. Veraswami, whose failure to get into the degenerate European Club becomes the mechanism of his destruction by his political enemies.

There is no question about Orwell's fundamental socialism, just as there is no question about his fundamental atheism. But those who know his work are aware that he did not automatically accept the socialist party line, and that he started to break with many British Socialists in the thirties when he concluded that the Russians were actually helping the Fascists in Spain. By the time he published *Animal Farm,* in the mid-forties, he was one of Britain's leading anti-Soviet writers, perceiving Russia to be as much an enemy of socialism as the Nazis had been. Frequently overlooked, however, is the fact that Orwell had another source of disagreement with his fellow Socialists, and although it attracted less attention than the flap over Russia, it was much more fundamental.

Orwell claimed that every time he wrote something positive about nature he got dozens of letters from angry Socialists calling him a bourgeois sentimentalist and saying that he ought to have more faith in progress and the ability of the machine to liberate the workers from oppression. In his essay on the common toad he wrote sarcastically, almost desperately:

> Is it wicked to take a pleasure in spring . . . is it politically reprehensible, while we are all groaning, or at any rate ought to be groaning, under the shackles of the capitalist system, to point out that life is frequently more worth living because of a blackbird's song, a yellow elm tree in October, or some other natural phenomenon which does not cost money and does not have what the editors of left-wing newspapers call a class angle?

And he went on to ask the question,

> If a man cannot enjoy the return of spring, why should he be happy in a labour-saving Utopia?

What Orwell understood, what gave him one of the greatest claims to the title of prophet, and what makes his prophecy useful and valid today—the idea he developed but never had time to finish and polish—is that neither of the great contemporary social theory/ systems, socialism and capitalism, has been able satisfactorily to encompass both human relationships *and* the environment. Socialism was his personal choice, but he knew that socialism, as practiced in his day in the communist states, did not have room for both people and their nonpolitical world of nature and agriculture. Since Orwell's death, we have become increasingly aware of the extent to which Soviet state planners ignored the role of nature in human affairs, especially in agriculture. An analogous statement, could be made about the influence of capitalist forces on Western farms and other Western environments. In both cases, the cost of leaving nature itself out of consideration has been incalculable.

In *The Road to Wigan Pier,* Orwell says repeatedly that machine civilization is here to stay, and that as long as we must live in such a machine-dominated, beehive society it is better that it be a Socialist than a Fascist one. But he also says:

> Once Socialism is . . . established, those who can see through the swindle of "progress" will probably find themselves resisting. . . . In the machine-world they have got to be a sort of permanent opposition.

This permanent opposition was not likely to be a majority.

> There are now millions of people . . . to whom the blaring of a radio is not only a more acceptable but a more *normal* background to their thoughts than the lowing of cattle or the song of birds. The mechanization of the world could never proceed very far while taste . . . remained uncorrupted, because in that case most of the products of the machine would be simply unwanted. In a healthy world there would be no demand for tinned food, aspirins, gramophones, gaspipe chairs, machine guns, daily newspapers, telephones, motor-cars, etc., etc.; and . . . there would be a constant demand for the things the machine cannot produce. But meanwhile the machine is here, and its corrupting effects are almost irresistible.

For Orwell, the first task was to establish socialism, then, equally important, to "humanize" it, as he put it, to arrange the relationship between people and their natural environment in a way that permitted the greatest number to maximize their humanity. He didn't have any practical suggestions about how to do this, and although he recognized that there are instabilities and self-destruct mechanisms built into the bureaucracy of machine society—what he called in his 1943 essay on "Poetry and the Microphone," the "loose ends and forgotten corners" in the "machine of government"—he did not develop the idea. Orwell never realized, as many scientists now do, that similar instabilities and limits exist in technology itself, especially in its interactions with nature, and that these instabilities and limits may be themselves capable of controlling the worst evils of machine life.

If Orwell had lived long enough to realize this, if he had lived into the age of nuclear power plant cancellations and the first stirrings of ecological agriculture, he might have explored the implications and found a measure of comfort in them. He might have realized that nature, although it can be damaged and mutilated, is not so easy to bring under control—that a world with a glass roof over it and concrete and plastic everywhere, although it may look possible on paper, is highly unstable and cannot work for very long. He might have seen that the Wellsian fantasies of technological control are just that, fantasies. And if he had reached that point, he might even have begun to rethink some of his ideas about the possibilities of controlling human beings. But he didn't live past the final flush of technological intoxication. He died when people still spoke of "atoms for peace." In this instance at least, his prophecy did not take him far enough.

So Orwell had two visions of utopia: one a vision of a world in which nature is cherished and improved by a gentle and caring human civilization, and the other a vision of a world in which people treat each other decently and fairly, without exploitation. Ultimately, as we see in *The Road to Wigan Pier,* the two visions came together in the picture of a "simpler," "harder," predominantly agricultural way of life in which the machine is present but under human control and "'progress' is not definable as making the world safe for little fat men." A world in which progress, itself, is not a form of exploitation.

But Orwell was the supreme realist of our time, and he knew that neither of his visions were close to attainment. What then do we do in the meantime? He gave his answer in the concluding paragraph of his essay on the common toad.

> At any rate, spring is here, even in London N.1, and they can't stop you enjoying it. This is a satisfying reflection. How many a time have I stood watching the toads mating, or a pair of hares having a boxing match in the young corn, and thought of all the important persons who would stop me enjoying this if they could. But luckily they can't. So long as you are not actually ill, hungry, or immured in a prison or holiday camp, spring is still spring. The atom bombs are piling up in the factories, the police are prowling through the cities, the lies are streaming from the loudspeakers, but the earth is still going round the sun, and neither the dictators nor the bureaucrats, deeply as they disapprove of the process, are able to prevent it.

This is prophecy in the classical style of Isaiah—warnings of a coming doom caused by collective moral failure always alternate with messages of consolation. If we ignore the guideposts and danger signs that are displayed by nature for anyone to read, our society will not long survive, and those who are reading the signs and shouting the warnings will be swept away with the rest. But in the midst of chaos there is still spring, if we care to notice it, and with spring the continuing hope that our prophets will at last be heard.

The Rain Forest
of Selborne

The fourth most frequently published book in the English language is by modern standards of popular entertainment one of the dullest. Unlike television commercials, in which images flicker on and off the screen every few seconds, the scenes in this book stay the same page after page. A cameraman filming *The Natural History of Selborne*—if any producer were crazy enough to try to turn that book into a television program—would merely have to aim the camera, turn it on, and go to lunch, returning only to reload film. No need to swivel the camera to follow the action; there is no action in *The Natural History of Selborne*. Nor are there any dismembered corpses, love polygons, towering ambitions, or anything that could be described as exotic or glamorously foreign, not even a jaguar or a tree fern.

The book was first published in 1788–89. As its title makes plain, it is an account of the natural history of the land in and around Selborne, a village in Hampshire, in south central England. Its author was Gilbert White, a lifelong bachelor clergyman in the district, who remained a lowly curate throughout his career, never attaining the position of vicar held by his grandfather, although he lived in his grandfather's house, "The Wakes," after the vicar's death. The format of the book is simple: it comprises a collection of 110 descriptive, anecdotal letters written by White to two acquaintances during the years 1767 to 1787.

At the time of this writing, there have been between two hundred and three hundred editions and translations of *The Natural History of Selborne,* an average of more than one per year for two centuries, and it is still going strong. If the book were an agricultural method or a managed forest, I suppose we would call it "sustainable." As an author, I have often wondered what it is that has kept *Selborne* alive all this time. Here is a typical passage, taken from the fortieth letter to Thomas Pennant.

> The fly-catcher is of all our summer birds the most mute and the most familiar; it also appears the last of any. It builds in a vine, or a sweetbriar, against the wall of a house, or in the hole of a wall, or on the end of a beam or plate, and often close to the post of a door where people are going in and out all day long. The bird does not make the least pretension to song, but uses a little inward wailing note when it thinks its young in danger from cats or other annoyances: it breeds but once, and retires early.

What is the appeal of this sort of passage (assuming that it appeals to you)? The prose is very good, but no better than much other prose written in the eighteenth century. The information is mostly accurate—although spotted flycatchers do have a faint song—but modern birding guides are more useful. Let's try another passage, this one from letter twenty-nine to Daines Barrington.

> In heavy fogs, on elevated situations especially, trees are perfect alembics: and no one that has not attended to such matters can imagine how much water one tree will distil in a night's time by condensing the vapour, which trickles down the twigs and boughs, so as to make the ground below quite in a float. In Newton-Lane, in October 1775, on a misty day, a particular oak in leaf dropped so fast that the cart-way stood in puddles and the ruts ran with water, though the ground in general was dusty.

What could be more static, less "newsworthy," than an account of a tree dripping water from its leaves and branches in a fog? Note that there is no poetry here, no images of ghostly fog, glittering branches, and mysterious, shimmering puddles—just one rather matter-of-fact metaphor, the tree as alembic, or distilling apparatus. Yet the enduring fascination of Gilbert White's book is plain to see in this short paragraph, especially to the jaded, overtitillated, adventure-soaked eyes of a typical circummillennial American. Despite their prosaic nature, I can't read White's words without a thrill of excitement and pure pleasure: how thrilling to be connected to a particular oak tree in a particular lane in Hampshire in October of 1775. How thrilling to be connected, attached, in contact with, anything real and straightforward in these days of images on a flat screen and deception as a way of life.

White describes a time and place in which natural history and human history were neither separated nor separable, when nature didn't have to be exotic or even beautiful to touch the heart. Consider these words from the twenty-eighth letter to Daines Barrington.

> It is the hardest thing in the world to shake off superstitious prejudices: they are sucked in as it were with our mother's milk. . . .
>
> In a farm-yard near the middle of this village stands, at this day, a row of pollar-dashes, which, by the seams and long cicatrices down their sides, manifestly show that, in former times, they have been cleft asunder. These trees, when young and flexible, were severed and held open by wedges, while ruptured children, stripped naked, were pushed through their apertures, under a persuasion that, by such a process, the poor babes would be cured of their infirmity. As soon as the operation was over, the tree, in the suffering part, was plastered with loam, and carefully swathed up. If the parts coalesced and soldered together, as usually fell out, where the feat was performed with any adroitness at all, the party was cured; but where the cleft continued to gape, the operation, it was supposed, would prove ineffectual. Having occasion to enlarge my garden not long since, I cut down two or three such trees, one of which did not grow together.

The point I am making with this quotation is not that people who are close to nature are superstitious and cruel, if indeed that is true. What I notice is that to Gilbert White the agony of "the poor babes" and the hurt of the trees in their "suffering part" are fused together in the same memory. When White described nature, he was also describing himself and the people who lived in his parish. Perhaps that is why he never traveled to view nature in distant places, even though he had the means and the interest. Maybe he thought he wouldn't be able to comprehend the plants, animals, soils, and weather in a place where he didn't live. Maybe he couldn't bear the idea of leaving Selborne. Maybe he thought foreign travel would be too exciting. This is hard for us to comprehend: our society knows how to repair ruptures, but we find it both difficult and unappealing to lead a wholesome, contented life at home. Rest for us, a vacation, means going away.

Even more striking than the enduring success of Gilbert White's local natural history is its paucity of successful imitators. Because there are as many localities as there are places where people live, the possibilities for successors to *The Natural History of Selborne* are infinite. But the successors—and there have been many of them—have rarely become popular. It is as if the English-speaking world needed only one book about the natural history of home.

The main reason for this limited interest in local natural history is that even as White was writing his last letters to Daines Barrington, the world was changing. In a few short decades, the first of a new kind of natural history book had appeared: Charles Waterton's *Wanderings in South America,* later known familiarly as "Waterton's Wanderings," which told of toucans, poison arrows, gigantic snakes swallowing deer, vampire bats, jaguars, and deadly fevers. There was still *Selborne* for nostalgia, but one *Selborne* was quite enough. A restless, rootless, exploring, dominating society wanted nature books to match its mood.

This mood is still dominant in the twentieth century in America, where it has been celebrated first and foremost by William Beebe, whose wonderful books transported bank tellers, dentists, and housewives into tropical jungles where army ants and bushmasters were never more than a few pages away. Beebe's earliest jungle book, *Jungle Peace,* published in 1918, was reviewed by Theodore Roosevelt in *The New York Times Review of Books,* a review that was subsequently included as the Foreword. Roosevelt praised Beebe and Waterton as being not only field naturalists and men of letters, but "men of action [who] have described for us the magic and interest, the terror and beauty of the far-off wilds where nature gives peace to bold souls and inspires terror in the mind."

In the second half of the twentieth century, the natural history books about far-away places have continued to appear and have been supplemented by televised nature documentaries, almost all of which are Waterton warmed over—glamorous accounts of bizarre life forms somewhere else. TV nature is "world-class" nature: all whales, warthogs, and wombats.

So it is with the conservation movement. Endangered tropical rain forests get a much better American press than endangered forests in New Jersey and Ontario or endangered prairies in Kansas or endangered family farms in Minnesota. True, the world's tropical rain forests are gravely threatened and are in desperate need of conservation. They harbor the bulk of the earth's species of plants and animals, which is very important. But beware of the numbers game. The species of the tropical rain forest will not help us restock devastated ecosystems if they occur in Oregon or southern Africa or Sweden or Chile or most places in the world where people live. Conservation has to start at home, where we know, or ought to know, the problems, and where we are most likely to understand the opportunities and limitations of our solutions.

Gilbert White was not a conservationist; there were no shopping malls or oil refineries in Hampshire in 1788. Today, when conservation is imperative everywhere, when the environment of everyone's home is endangered, the intimate closeness to local nature which is the common ingredient of all good conservation is missing. The homeless, domineering human spirit that has caused all modern environmental problems also, sadly, underlies our preoccupation with exotic nature. I am left wondering, how is sustainable conservation going to be possible if conservation has to be exciting enough for television? If Gilbert White were starting fresh today in his still-beautiful village, without the benefit of a two-century reputation, he would certainly have a conservation message, but he would have trouble finding a publisher—unless he were willing to change the title of his book to *The Rain Forest of Selborne.*

Desert Life

Occasionally, for the best and worst of motives, we feel compelled to leave home. When we do, we generally find that however difficult it is to get along with our own familiar if misunderstood environment, it is far harder to exist comfortably and unobtrusively in somebody else's.

The arrival of American troops in the Middle East in the summer of 1990 set loose a sandstorm of media stories about the intolerably harsh conditions in the Arabian Desert. From the wastes of An Nafūd and Ad Dibdibdah in the north near Iraq to the vast Ar Rab' al Khālī, or Empty Quarter, in the south next to Yemen and Oman, we saw the sand and stones stretching on endlessly from horizon to horizon in a lifeless, waterless panorama. Soldiers had to drink several gallons a day to prevent dehydration—what an impossible environment for living creatures!

These media stories brought to mind the fable of the town mouse and the country mouse. You probably remember it: the town mouse, visiting his country cousin, is appalled at the rustic conditions and simple fare, so he invites his cousin to see how sweet life can be in the urban fast lane. While they are dining off sumptuous leftovers in a regal dining room, a ferocious watchdog appears (maybe a cat in your version) and the two mice have to run for their lives. The town mouse takes it in his stride, but the country mouse, frightened out of his wits, leaves the roast pheasant and aged camembert and rushes

34

back to the security of stale crusts of black bread and rinds of rough cheese shared with a kindly farmer. Put another way, my part of the Middle Atlantic states can be a terribly harsh environment to anyone who insists on roller skating down the New Jersey Turnpike or surfing in a cotton bathing suit in February. Any environment is harsh if you aren't adapted to it.

There is plenty of life in most deserts, including such storied ones as the Arabian Desert and the Sahara. Closer to home, the arid Southwest of the United States teems with plants and animals—for example, more kinds of wildflowers and bees live there than in any other region of the country. For them the desert is home; many cannot survive anywhere else. We have to assume that the desert is no harsher for them than Bermuda in the springtime would be for us.

In the warm deserts of the world there are two problems that must be solved by all inhabitants, the heat and the dryness. Coping with either one by itself often makes the other one worse: a plant or animal that cools itself by evaporating water soon runs out of water and dies, so except for a few special cases they do not cool themselves that way. Instead, organisms usually attack the problems of heat and dryness simultaneously, and this limits the number of possible solutions. That is why desert plants and animals often have similar adaptations.

One similarity of desert plants and insects is in their outer coverings. The epidermis of insects and the epidermis of plants both have an outer waxy layer with the wax molecules oriented perpendicular to the surface. This waxy coating prevents evaporative water loss but it also acts as a barrier to oxygen and carbon dioxide, which must be able to get in or out as needed. The solution to this secondary problem is also the same for plants and insects—the epidermis is provided with small holes that can be opened or closed according to need and outside conditions.

Tolerance of heat and water loss is another adaptation that many desert creatures share. The small plant *Tidestromia oblongifolia,* which lives at the bottom of Death Valley, can photosynthesize normally at temperatures of at least 122 degrees Fahrenheit, when many plants would be wilted and inactive. Special enzymes and deep roots probably have something to do with its hardiness. The creosote bush of our southwestern deserts can tolerate the loss of much of its

internal water without dying; it only hangs on more tightly to the
water that is left. The larvae of midges that live in temporary ponds
in West Africa dry up with the ponds after the seasonal rains end. In a
desiccated state they can survive temperatures as high as 216 degrees
Fahrenheit for short spans of time.

Other plants and animals hide from the desert sun. Some plants
"hide" by shutting down metabolic activity except during early
morning and late afternoon hours or during cooler and moister times
of year. This way they can keep their leaf pores closed during the
hottest times and prevent water loss. A few plants such as cactuses
and euphorbias have eliminated their leaves, and by pointing the
short diameters of their green stems at the sun offer little surface area
for heat uptake—another form of hiding. Desert lichens hide just
underneath the outer crusts of rocks and pebbles, surviving in the
minimal shade despite the blistering heat. Animals hide in the shade
of bushes and rocks or dig burrows which they leave only at night.

Desert plants are adept at extracting water from dry soils, spreading fine rootlets on the undersurfaces of stones where the soil moisture is highest. Some burrow-dwelling spiders can suck the film of
moisture from the sides of their burrows, even against a tremendous
pressure gradient. Kangaroo rats are among those desert mammals
that never drink; the digestion of their food provides all the water
they need. Many desert plants absorb the dew that collects on their
leaves at sunrise—the importance of dew has been known by Mid-
eastern peoples for thousands of years.

Of all desert animals the camel is justly the most famous. Its
ability to thrive in desert conditions is legendary, and for once the
legends are correct. One typical feat was described by Professor
Theodore Monod, of Dakar, who between December 12, 1954 and
January 2, 1955 traveled in a caravan on camelback across the Empty
Quarter of the northwestern Sahara (not the Empty Quarter of the
Arabian Desert), twenty-one days and nearly six hundred miles
without water for the camels. We understand a good deal about how
the camel does it thanks to studies in the forties and fifties, especially
by the physiologists Knut and Bodil Schmidt-Nielsen.

First, what camels do not do. They do not store water—not in their
humps, not in their stomachs, not anywhere. This notion probably

grew out of observations that camels can drink prodigious amounts of water, around thirty percent of their normal body weight in one session. The Schmidt–Nielsens observed a camel that drank twenty-five gallons of water in a few minutes and then took another twenty-four gallons later in the day, but this and other such drinking sprees only serve to make up for water already lost; once camels have reached their usual weight they stop drinking. Nor can camels tolerate excessively high body temperatures. It seems likely from circumstantial evidence that the lethal temperatures for camels and humans are about the same, 105 to 108 degrees Fahrenheit.

Camels survive the heat and dryness of the desert by a number of adaptations, no one of which might suffice to keep them alive. They can tolerate extremes of dehydration, losing water equivalent to at least 27 percent of their body weight without suffering permanent damage, compared to the 12 to 14 percent limit for other mammals. Most of their subcutaneous fat is in the hump, leaving the rest of the skin without an insulating layer and therefore capable of losing heat rapidly. Camels sit in the same spot all day, huddled together in groups with their legs folded under them; all of these behaviors reduce heat gain in an environment where there is no shade. They can excrete highly concentrated urine and very dry feces, greatly reducing water loss. They also seem to be good at recycling broken-down protein, which frees them from the necessity of getting rid of much urea (the main protein waste product) in urine. And camels have other ways, both known and certainly unknown, of making the desert a pleasant place in which to live.

The photographs of American soldiers stripped to the waist in the Arabian September sun reminded me that people also have worked out ways of living in the desert—shedding clothing is not one of them. Here is what Knut Schmidt–Nielsen had to say about clothing in his book *Desert Animals:*

> It may seem paradoxical that clothing should be an advantage in hot surroundings. . . [but] I have found that the tendency to shed clothing is reversed when the temperatures are very high. When the Sahara Desert really heated up in June, it seemed unbearable to walk around in the very light clothing we had been using earlier in the

year. We switched to long baggy Arab trousers and long-sleeved shirts. In the cool morning . . . [we had our] shirt sleeves rolled up, but as the baking heat of the mid-day approached the bare arms felt uncomfortable.

The same woolen garments are worn winter and summer by Arabian Bedouins: in summer they offer protection against heat, in winter they ward off the desert cold. The robes of Bedouin men are white, those of women, who tend the goats, are black. It was shown by Amiram Shkolnik and his coworkers at Tel Aviv University that although the black robes gain two to three times as much heat as white, this heat never reaches the skin. It is carried away by convection currents inside the loose robes, a kind of chimney effect.

Some of the most remarkable human adaptations to desert life have been in agriculture. The Indians of northern Mexico's and southern Arizona's Sonoran Desert modeled their agriculture on the desert around them. So successful were they that the early European explorers never realized they were walking through Indian farms and not the wild desert. Even the nomadic Bedouins "cultivate" the desert in subtle ways. The nineteenth-century explorer Charles Doughty, whose book *Travels in Arabia Deserta* inspired T. E. Lawrence, described evenings around the Bedouin campfires:

> The younger or meanest of the company, who is sitting or leaning on his elbow or lies next the faggot, will indolently reach back his hand from time to time for more dry rimth, to cast on the fire, and other sweet resinous twigs, till the flaming light leaps up again in the vast uncheerful darkness. The nomads will not burn the good pasture bushes, *gussha,* even in their enemies' country. . . . I have sometimes unwittingly offended them, until I knew the plants, plucking up and giving to the flames some which grew in the soil nigh my hand; then children and women and the men of little understanding blamed me, and said wondering, "It was an heathenish deed."

Some of the finest desert agriculture of all times was practiced by the Nabateans, who lived in Israel's Negev Desert two thousand years ago. Growing luxuriant crops in places that receive three to four inches of rain in a good year, the Nabateans worked miracles.

They placed their farms in the valleys of the hilly desert, lining the hills with runoff channels to convey every possible drop of rain to the farms below. The runoff channels terminated in a complex water distribution network of irrigation canals and weirs.

Two of these Nabatean farms, one at Shivta and one at Avdat, were reconstructed by Michael Evenari, late professor of botany at Jerusalem's Hebrew University, along with his wife, Liesel, and many colleagues. Through a combination of brilliant archaeological and botanical studies, Evenari rediscovered the extraordinary practices of the Nabatean farmers, including the probable reason for the thousands of mysterious piles and rows of stones that they left on the hillsides. Previous investigators had focussed on the piles and rows, with no success. By experimenting, Evenari found that raking the stones out of the desert soil greatly increased the flow of rainwater to the farms below. The much-studied stone piles had had no special function; they were trash heaps.

Using only Nabatean water sources and plowing with camels, the Evenari team was able to grow peaches, pistachios, apricots, apples, almonds, cherries, and grapes, in addition to field crops, vegetables, medicinal and pasture plants—the ultimate demonstration of the Nabateans' genius. Three thousand years earlier, before the Nabateans arrived, the Israelites, wandering a few miles from Avdat, complained to Moses: "Why did you make us leave Egypt to bring us to this wretched place, a place with no grain or figs or vines or pomegranates? There is not even water to drink!" (Num. 20:5)

Like most other places on this living planet, the desert is what one makes of it. Making something of it takes time. For countless ages people have found the time. Michael Evenari, Leslie Shanan, and Naphtali Tadmor explain why in their book, *The Negev:*

> It is night in the desert . . . the dark brings relief to the aching body, parched by heat. Enormous stars hang low over the dustless, dry atmosphere. Nothing stirs . . . Here are no whispering trees, no veils of cloud, no grassy meadows for the disporting of gods and goddesses. Here is that "voice of stillness" in which God spoke to his prophet Elijah. Here, in this silent world, not soft, not gentle, in this awesome boundlessness of time and space, one . . . stands alone, touched by Infinity.

When we travel to untamed places, it behooves us not to dismiss them as bizarre, to condemn them for their strangeness and wildness, or to try and bludgeon them into a complacent submissiveness. Better to recognize the life they contain and to find in them the inevitable elements that speak to all human souls. Then, perhaps, we will be less inclined to cause needless destruction and to take an angry departure. Then, perhaps, we will recognize the opportunity that wilderness offers to let us reach and develop the finer parts of our own natures.

A Turtle Named Mack

Early October of '91 was warm and frost-free in central New Jersey, just right for a family weekend walk along the towpath of the Delaware-Raritan Canal. Kate, the oldest child, was away at college, but everyone else was there. The three Js, Jane, age fifteen, Jon, who is nine, and my wife, Joan, went on ahead. Sam and I hung back. Sam, six years old and youngest, was digging a pit trap for beetles in the middle of the narrow towpath, while I was cautiously investigating the water at the edge of the canal. I was the only one with a secret agenda.

For nearly a year Sam had been agitating for a turtle, and one had been promised him. Friends whose life styles brought them into occasional social contact with turtles were alerted; we went on turtle-catching expeditions. No turtles were forthcoming. A so-called turtle expert, I was embarrassed. Sam was outwardly patient, but there were early signs of loss of faith.

So I was quietly and clandestinely looking for turtles, hoping to surprise one basking, eyes closed, in the autumn sunshine, next to the bank. It had to be right next to the bank because I had left the net at home, not wanting to dash Sam's hopes with another turtleless hunt. There were a few nearby plops as wary painted turtles slipped off logs into the water at my approach, and, once, a musk turtle startled me by jumping off an overhanging tree branch into the water about four feet from where I stood. Musk turtles are the only turtles I know of that climp up into trees to sun themselves. It isn't proper behavior for turtles, but they do it anyway.

The sun was dropping low in the sky; the air was getting chilly; there didn't seem to be any chance of catching a turtle until spring; I was depressed. As I broodingly stared into the canal, looking at a smooth brown stone on the shallow bottom, I gradually became aware that I had put the stone into the wrong mental category. The stone was slowly cruising along the bottom, neck outstretched, poking its head into the ooze in a search for food. The stone was a musk turtle, and it didn't know I was watching. But it was just out of arm's reach from the bank.

This was the moment of decision. I hastily reviewed some pros and cons. Pro: it was a musk turtle, not an endangered species, small and very aquatic (except when climbing trees), therefore adaptable to life in a good-sized tank. Also, Sam wanted and had been promised a turtle. Con: musk turtles are smelly, irascible, and compared with other kinds of turtles extremely ugly. Worse yet, I would have to wade into the canal after it and there was no time to take off my shoes. The turtle was moving away. I thought of Hamlet and all the bad things that happened to him because he didn't make up his mind, and stepped carefully over the poison ivy at the edge of the bank into the water and down into the foot of muck that concealed the true bottom of the canal.

Two minutes later, I clambered back up to the towpath where Sam, oblivious, was putting the finishing touches on his pit trap. The turtle was a small female. She had her mouth open ready to bite and stank horribly. Sam was delighted. Holding the turtle carefully by the sides of her shell, he ran off to find the others. As we walked back together, Sam magnanimously allowing his brother to hold the turtle and I trying to keep my hands downwind, we discussed possible names. We all had suggestions, especially Jane, but Sam was not interested. The turtle, he announced, was named Mack; after the hero of Dr. Seuss's great morality tale, *Yertle the Turtle*. And this brings me in a roundabout way to the point of *my* tale.

If you have children or grandchildren, or if you were a small child yourself any time after 1950, you ought to know the story of *Yertle the Turtle*. I have it in front of me and will paraphrase it to jog your memory. Yertle is king of the pond on the Island of Salamasond.

> A nice little pond. It was clean. It was neat.
> The water was warm. There was plenty to eat.
> The turtles had everything turtles might need.
> And they were all happy. Quite happy indeed.

Yertle, however, needs more scope. The domestic scene sickens him; he yearns to be a world-class king. Summoning his subjects, he orders them to make a living throne out of turtles, a "nine-turtle stack." Clambering up to the top of the heap, he finds his horizon expanded to include a cow, a mule, a blueberry bush, a house, and a cat. " 'I'm Yertle the Turtle!' " he proclaims, " 'Oh marvellous me! For I am the ruler of all that I see!' "

After a few hours, the king is respectfully interrupted in his gloating by Mack, the bottom turtle in the stack, who is feeling the weight. Angrily silencing him, Yertle adds a few hundred turtles to the precarious pile, making it high enough for him to see forty miles. Again, Mack, speaking for the lower echelons, meekly describes his pain and hunger, and is answered from above.

> "You hush up your mouth!" howled the mighty King Yertle.
> "You've no right to talk to the world's highest turtle.
> I rule from the clouds! Over land! Over sea!
> There's nothing, no NOTHING, that's higher than me!"

Night falls. A crescent moon rises. Yertle, infuriated because the celestial body dares to be higher than he is, calls for another 5,607 turtles. That does it for Mack. Deliberately burping, he sends a seismic tremor through the pile of turtles, unthroning the Turtle King and ending his oppressive rule.

> And today the great Yertle, that Marvellous he,
> Is King of the Mud. That is all he can see.
> And the turtles, of course . . . all the turtles are free
> As turtles and, maybe, all creatures should be.

The overt message is thus one of freedom, but there is much more to it than that. A common theme runs through many of the works of Dr. Seuss, like a stream flowing through a checkered landscape, binding together fields, villages, woods, marshes, and hillsides. The theme is that life, as it should be, has an intrinsic order and pattern in which all plants and animals, including us, have a place, a piece of the pattern. Each place is defined; boundaries are established, but the area within the boundaries is always ample because the definition of the piece of the pattern is the same as the description of the natural limits and capabilities of its particular occupants.

The message that an order, a pattern, exists in the world is immensely reassuring to children, and essential, I believe, for their healthy development. Imagine yourself placed in a totally alien landscape throbbing with strange and sometimes menacing activity and dotted with important-looking signs written in a language that you cannot read. How grateful you would be to a kindly, bilingual stranger who could teach you to read the signs and get your bearings. There are landmarks that show us the pathways and limits of our place—all species have them. But humans, far more than any other species, must actively acquire knowledge of these landmarks through learning. And only humans are free to ignore the landmarks, to pretend that they are not there. It was the lifework of Dr. Seuss to teach these landmarks to children, and in so doing to show them and us how to identify both the safe roads and the boundaries that must not be passed in a living world of wonders and pitfalls.

The wonders and pitfalls are described in every book that Dr. Seuss wrote, making the order and pattern vividly clear to the slowest child and dullest adult. *And To Think That I Saw It On Mulberry Street* is the story of a little boy named Marco who enlivens his day by imagining fabulous sights that he sees on his way home. But when his father presses him to describe what he saw, the bejewelled rajah riding on an elephant flanked by two giraffes, all harnessed to a bandwagon complete with brass band and motorcycle escort, turn back into "a plain horse and wagon on Mulberry Street." It is all right—even necessary—to use our imagination to transcend the limits of our world, but we mustn't forget the difference between fantasy and reality. There is a time to come back to earth.

Imagination is not the only source of extraordinary happenings. They occur naturally, and when they do, sometimes the only acceptable course of action is to adapt and make the best of them, as we learn in *The 500 Hats of Bartholomew Cubbins*. Young Bartholomew Cubbins removes his peasant's feathered cap as King Derwin rides by, but another cap mysteriously replaces it. The king has Bartholomew arrested for lese majesty as cap after cap pops into place every time he bares his head. What makes it worse, after a while each new cap is more elaborate than its predecessor. Escaping execution on a technicality (it is forbidden to behead someone who is wearing a hat), Bartholomew is saved from further threats by the dazzling splendor of his five-hundredth hat. The king's rage evaporates, he wants the hat and offers to buy it, and when the dazed Bartholomew accepts the offer and the king himself lifts off the hat with trembling hands, the hat plague is ended as suddenly and inexplicably as it began.

Innocent fantasies and inexplicable events are part of normal existence—they illuminate the topography and ultimate boundaries of our place without changing them. Deliberate efforts to alter the boundaries are another matter; they always lead to trouble. In *Bartholomew and the Oobleck,* Bartholomew Cubbins returns in the greatest of all of Dr. Seuss's object lessons for human survival. King Derwin, it seems, is bored with the four things that always come down from his sky: " 'This snow! This fog! This sunshine! This rain!' " He orders Bartholomew to blow the secret whistle that summons the royal magicians.

" 'Your *magicians,* your Majesty?' Bartholomew shivered. 'Oh, no, Your Majesty! Don't call *them!'* "

But the king insists and the magicians appear. Ordered to make something new that falls from the sky, they hesitate only a moment and then unanimously decide on oobleck. The magicians spend the night in incantations and the oobleck arrives at dawn. It is green, blobby, and has the consistency of fresh-mixed epoxy glue. Everything it touches is stuck fast: farmers to their plows, birds on their nests, the Royal Fiddlers to their fiddles, the king to his throne, the alarm bell to its clapper—even the door to the magicians' cave is sealed.

King Derwin, cut off from his magicians, desperately tries to remember and concoct magic words that will halt the descent of the oobleck. Nothing works. The oobleck, now pelting down in basketball-sized globs, penetrates every crack and crevice in the kingdom. Then Bartholomew loses all patience.

" 'Don't waste your time saying foolish *magic* words. YOU ought to be saying some plain *simple* words! . . . this is all your fault! Now, the least you can do is say the simple words, "I'm sorry." ' "

Derwin replies. " 'Kings *never* say "I'm sorry!" And I am the mightiest king in the world!' "

" 'You may be a mighty king,' " Bartholomew answers, " 'But you're sitting in oobleck up to your chin. And so is everyone else in your land.' "

The force of Bartholomew's argument and moral stance prevail, the king repents and apologizes, the sun appears, and the oobleck melts away. Never again will the king try to change the givens that at once delineate his world and define his rightful place in it.

In other books by Dr. Seuss, other warnings are given; I will mention only two. *The Lorax* makes it clear that a world made needlessly desolate will support neither its accustomed animals and plants nor the people who caused the desolation. And *The King's Stilts* introduces another and very modern ecological teaching: humans can, slowly and with care, modify their environments in beneficial ways, but it takes a vigorous, perpetual exercise of knowledge and skill to maintain these changes.

Theodor S. Geisel, Dr. Seuss, died on September 24, 1991. His books remain, enduring testimony to a long life nobly spent in helping children and their parents find their way safely around the intricate landscape of our earthly place.

And what of Mack? She seems content enough in her tank, spending much of the day under the rock bridge we made out of New Jersey Highlands stones held together with aquarium cement. Every now and then she emerges to pounce on an earthworm or other morsel that Sam drops in for her benefit. But it isn't really her world—for one thing, her artificial habitat has no male musk turtles around and no places to lay eggs. In the spring, after the forsythias have bloomed and the willows have leafed out but before the musk turtle nesting season is underway, when she is eating well and has shaken off her winter lethargy, I will explain to Sam that it is time to take her back to her home. Sam is a reasonable fellow and is likely to agree. So we will take Mack to the Delaware-Raritan Canal, her place, and I will let Sam put her carefully in the branches of a tree overhanging the water, and we will watch her dive off into the canal and disappear into the ooze.

II

Off Course

Now the sneaking serpent walks
In mild humility,
And the just man rages in the wilds
Where lions roam.

WILLIAM BLAKE
The Marriage of Heaven and Hell

The Overmanaged
Society

We shall have to refer it right and left; and when we refer it
anywhere, then you'll have to look it up. When it comes back to us at
any time, then you had better look *us* up. When it sticks anywhere,
you'll have to try to give it a jog. . . . Try the thing and see how you
like it. It will be in your power to give it up at any time if you don't
like it. You had better take a lot of forms away with you. Give him a
lot of forms!

CHARLES DICKENS, Little Dorrit

There is an extraordinary proliferation of managers in our society,
an ever-increasing percentage of people who control events but do
not themselves produce anything real or useful. The problem of the
growth of management and its influence penetrates nearly every area
of modern life. Yet after decades of pervasive overmanagement,
comparatively few people understand it as a widespread destructive
force.

Overmanagement is a by-product of an exploitative age in which
the massive extraction and processing of natural resources have been
accompanied by the release of huge amounts of surplus wealth.
Managers feed on this wealth, dissipating it as management grows
and rendering it unavailable for future use, somewhat like the burning of waste gases at an oil refinery. Unlike gas flares, however, the
growth of management is uncontrolled: eventually it consumes and
extinguishes the power of the society that nurtured it, as resources
dwindle and wealth, wasted, declines. This has been our experience
in the last years of the twentieth century.

The abuse I describe is an excess of something that is perfectly all right, even necessary, in reasonable quantities. Management and managers are needed in modern society. If this is an indictment, it is an indictment of modern society, not management. Yet management, from being a service, can become a raison d'être and take on a life of its own. Swelling beyond all reasonable size, it appropriates and stifles the life of society; at this point it becomes utterly counter-productive and destructive. As a result, we stand at the beginning of a new societal conflict, the successor to the Marxist-capitalist debate: the struggle of the producers of goods and services against central-ized management.

The magnitude of the problem is so great that like the Grand Canyon, it offers only isolated, partial views. The best are provided by the federal government. As Paul Von Ward has estimated, in 1975 the cost of the government paperwork was already one hundred billion dollars, one-fifteenth of the total U.S. economy. It has gotten steadily worse since then. "Perhaps more than one-half of the overall resources consumed by the federal government are non-productive transfers of funds," Von Ward stated; "payments to people who produce little or nothing is simply another form of welfare."

The State Department is a case in point. In 1945, the *U.S. Government Manual* listed, in addition to the Secretary of State, 1 under secretary, 6 assistant secretaries, 14 directors, 10 deputy directors, and 58 chiefs. This apparatus supervised the activities of 315 foreign embassies, legations, consular offices, and missions. The country was at war. By 1975–76, State Department administrative positions listed in the manual included, in addition to the secretary, 3 am-bassadors-at-large, 1 deputy secretary, 3 under secretaries, 1 deputy under secretary, 13 assistant secretaries, 115 directors, and a host of deputy assistant secretaries and deputy directors. There was no longer room in the manual for chiefs. The number of foreign diplo-matic posts had declined to 274. The country was at peace. For 1990–91, the manual, which was the same size, was less useful because few of the positions listed in it fell below the rank of assis-tant secretary. There were 17 of them—and still 274 foreign posts.

The qualitative changes in the State Department are even more informative than the numerical totals. Most striking is the rise of administration at the expense of foreign affairs as the chief business of the State Department. In the 1975–76 manual, the Assistant Secretary for Administration was the seventh of his rank listed; the assistant secretaries for the geographic regions of the world took precedence. By 1990–91, the Assistant Secretary for Administration was first. Also, the Under Secretary for Management coordinated 6 bureaus, 5 offices, and an institute—each with its own subdivisions. The old functional clarity of the 1945 department, organized to reflect the geopolitical divisions of the world, has been lost, a victim of the explosive fragmentation of administrative categories, many with overlapping jurisdictions. More offices and administrative positions are created to coordinate the mess, and there emerge monstrosities such as "Vice Chairman and Executive Director of the Under Secretaries Committee Steering Group on Joint Cooperation Commissions." It is hardly surprising that in the last few decades much of our important foreign policy has been formulated outside the State Department.

Overmanagement is not confined to government. Hospitals present a similar picture of exploding management, as managers divert funds to hire more managers. Consider one 260 bed, voluntary, nonprofit hospital less than a hundred miles from New York City. Ten years ago, a typical department, obstetrics, had the following administrative chain of command in nursing: A Director of Nursing for the hospital, with one assistant, supervised two head obstetrical nurses (one for deliveries, one for postpartum care). Today, a Vice President for Inpatient Services supervises the Clinical Director of Nursing (in obstetrics), who in turn directs five nurse-managers. The number of deliveries per year had not changed. At the same hospital, seven vice presidents are in charge of fifty-five administrative units, most of them with directors and managers who supervise the real work of the hospital—which apart from the reams of new administrative paperwork is not so very different than it was ten, twenty, or thirty years ago.

In the same vein, a Canadian surgeon from a university hospital in Ontario wrote me following some televised comments that I made on overmanagement: "I came to the hospital in [——] about 35 years ago. At that time there was one Administrator, a Comptroller and a Head of Nurses as the bureaucracy. Now, at last count, that bureaucracy has swollen to at least 20. In the process the hospital has gone

from 330 beds to 230 beds!'' Here, without doubt, is a significant
cause of the astonishingly high cost of hospital care, along with the
deterioration of hospital service and its sluggish reponsiveness to the
needs of patients.

Why Management Expands

One of the first and most vivid descriptions of a bloated bureaucracy
is in Charles Dickens's *Little Dorritt,* first published in 1857. At that
time, the phenomenon of overmanagement was already well developed in government if not in the university or in business. Bureaucrats thrived, even though urban populations were smaller than today's, technology was less complex, and computers, copiers, and
fax machines were nonexistent. In *Little Dorrit,* Dickens invented a
government agency called the Circumlocution Office, whose function was "HOW NOT TO DO IT," a task achieved by the liberal use
of forms and interoffice referrals.

It was brilliant satire, but I don't think Dickens understood why
overmanagement came about. He never perceived that it was an
organic outgrowth of a power–worshipping society dedicated to a
belief in control.

In *Little Dorrit,* we read of the unhappy fate of an inventor stymied
by the Circumlocution Office. To Dickens, an inventor was, quite
simply, a noble person who ought not be thwarted in his work. What
he didn't realize was that invention and technological cleverness, in
the context of England's growing cities, were responsible for the
Circumlocution Office in the first place. The sudden access and
concentration of power brought on by industrial technology came
hand in hand with management. New kinds and levels of production
were inevitably accompanied by squadrons of people who produced
nothing in the way of tangible goods, but managed the factories,
distribution networks, and government taxation and regulatory offices that constituted the new production system. It took about three-
quarters of a century for this idea to occur to Lewis Mumford and
George Orwell, among others.

Orwell saw the connection quite clearly, although he died before
his own insights into technology and administration could mature. In

a book review written in 1945, he wrote: "The processes involved in making, say, an aeroplane are so complex as to be only possible in a planned, centralised society, with all the repressive apparatus that that implies." In other words, a technologically complex society requires a good deal of management. Orwell associated this "repressive apparatus" especially with the manufacture of expensive, sophisticated weapons systems, but I think we could say that a society that produces antibiotics and VCRs will also have to have more administrators than one that does not.

Nevertheless, the process has gotten out of hand. Having initially expanded in response to a real need to organize complex processes, management continues to expand according to its own self-generated imperatives. Like a cancer, it has become uncoupled from the organism that produced it. Moreover, it has spread beyond production and government services to nearly all walks of life: from hospitals to universities, the dead hand of the manager is increasingly felt.

Management spreads because its methods and output automatically create an environment conducive to its own increase. Each manager doing his or her job brings forth more and more managers, as C. Northcote Parkinson was one of the first to show in his grimly amusing *Parkinson's Law,* first published in 1957. The process need not be conspiratorial or even purposeful, which is what makes it so difficult to stop. It is a classical case of positive feedback, with several feedback loops involved.

There are two elements in the job of management that fuel this feedback. First and most obvious is the increasing habit of documenting everything. Anyone who has ever had the misfortune of compiling a university promotion packet or filling out tax forms knows that it is no longer God alone who records all our deeds and numbers the hairs of our heads.

The true function of this modern tendency to document everything is plain. It serves to provide a source of undemanding work for managers who might otherwise not be terribly busy—thus justifying the call for yet more managers and conferring a kind of spurious legitimacy on the perpetual growth that is inherent in modern administration. More important, compliance with administrative demands for ever more minute personal and other information reinforces the desired belief that the provider of the information is subordinate to the recipient.

Most of us in nonmanagerial positions have become so accustomed to this routine violation of privacy that we think of it as nothing more than an annoyance. It is easy to forget that just as in some code systems, the data being transferred are often themselves irrelevant, a blind. The real information is symbolic and resides in the act of transfer. This explains why the data are often filed away unread and why the weary respondents are asked for the same information over and over again. The fact that one has filled out a form is important; what is on the form is not.

Frequently, this paperwork is justified as part of a process of judgment—words such as "evaluation" and "accountability" are used. These terms further establish the idea that the administrative evaluators are fit to judge, even though this is often not the case because the administrators are at least once or twice removed from the work itself and are not intimately familiar with it. When managers use abstract nouns such as "excellence" and "vision," what is usually being conveyed is a feeling, not a meaning. They are establishing a dominance relationship, not describing realities of the work at hand.

The second way in which administration expands involves a far less subtle mechanism than the proliferation of paperwork. It is the direct appropriation of power through control of the money supply and of hiring and firing. We see this mechanism at its highest degree of refinement in the modern university, which has come to be dominated by priorities determined by a cash flow controlled by administrators. In the words of the great biochemist Erwin Chargaff, universities "have been turned into huge corporations whose only business is to lose money." But universities are not the only places where management has diverted and augmented the money supply to provide for its own continued growth at the expense of other parts of the institution. It happens in charities, businesses, hospitals and other quasi-public institutions, and the many branches of government.

What makes bureaucracies spiral out of control in their demand for institutional funds are the positive feedback loops that so many bureaucracies participate in and encourage. Universities again provide one of the best examples. In the decades immediately following the

Second World War, university administrators in the United States discovered a vast new source of unregulated cash, the so-called overhead or "indirect costs" surcharge on research grants. Administrators, negotiating with their government counterparts, could set the overhead at incredible levels—figures in excess of fifty percent became common—and the incoming funds disappeared into an administrative black box. Pressure on scientists to obtain more grants increased steadily, more overhead dollars flowed in, more administrators were hired, more money was needed. A classic positive feedback loop thus became established: as the output of the system (grant overheads, patents, industrial research contracts) increases, it serves, through yet more powerful administrative pressures, to stimulate a further increase in output, and the loop, enhanced, begins again. Under the influence of administrative demands, the purposes of grants become irrelevant; what matters are the amounts of the grants with overhead and their likelihood of attracting more. All hiring and promotion of faculty members must be approved by the administration, while the hiring and promotion of administrators need the approval only of senior administrators and not of faculty. The positive feedback is augmented further. There is no negative feedback, or brake, built into the system. This system has changed the character of the faculty and the entire university. Other kinds of organizations, governmental and business, have similar kinds of positive feedback loops—in each case, managers have total control of the money flow and of hiring and firing.

The Consequences of Overmanagement

First, too much management causes *bad decisions*. As administrators become more numerous and as the power gulf between administrators and producers widens, more and more of the critical decisions for an institution are made only by administrators, on the basis of second- or third-hand information and in accordance with purely administrative priorities. The people who know are shut out of the decision-making process. The result is bad decisions, out-of-touch decisions.

David Bella, a professor of civil engineering at Oregon State University, has developed a general model which, by analysing the information feedback loops at different organizational levels, shows how only "favorable" or supportive information, information that confirms the rightness of the administration's decisions and actions, is allowed to move upward in the organization to decision-making levels. Bella sent his model to Nobel laureate physicist Richard Feynman, the leading investigator of the *Challenger* disaster, a disaster that was probably caused when unfavorable information (about the poor cold-tolerance of the shuttle's O-ring seals) failed to move upward to command levels of NASA. Feynman responded: "I read Table 2 and am amazed at the perfect prediction of the answers given at the public hearings. I didn't know that anybody understood these things so well and I hadn't realized that NASA was an example of a widespread phenomenon."

The failure of "unfavorable" information to move upward in the administrations of businesses, government agencies, and universities is the cause of a surprisingly large number of bad and costly decisions. Most are never brought to light, but a few of the most acutely horrendous ones have found their way into the newspapers. One example was given in a front page story of *The New York Times,* December 26, 1988, entitled "Nuclear Arms Industry Eroded As Science Lost Leading Role." Explaining the rundown equipment, shoddy management, and environmental problems at the Hanford, Washington, facility, Harold Agnew, a distinguished physicist, wrote: "The guys making the decisions don't understand the technical things any more. . . . You can't run the bomb factories with a bunch of lawyers and administrators."

In another front page article in *The New York Times,* dated January 16, 1989, reporter Keith Schneider, writing about the Savannah River nuclear weapons plant, said: "Internal memorandums prepared by DuPont, which built and operated the vast weapons plant for nearly four decades, show that company scientists amassed volumes of research on key weaknesses in equipment." Yet so effectively was this vital information buried by the corporate and governmental bureaucracies that even Westinghouse, which was scheduled to take over the plant from DuPont in April of that year, and which was certainly privy to all the top secret information about it, said that it knew nothing about the seriousness of the problems at Savannah River.

Education has been especially hard hit by overmanagement. In a *New York Times* opinion piece dated December 9, 1988 and entitled "To Revive Schools, Dump Bureaucrats," John E. Chubb a senior fellow at the Brookings Institution, said:

> New York City, as I discovered after a 35 minute phone call with nine different bureaucrats at the Board of Education, has 6,622 full time employees in its public school headquarters. That's one external administrator for every 150 students. [This does not include administrators located at the schools.] By comparison, the Roman Catholic Archdiocese of New York has so few employees in its headquarters that the first one I called simply offered to count them for me—30 central administrators; no more than one for every 4000 students.
>
> In a new study of more than 400 American high schools, a Stanford political scientist, Terry Moe, and I conclude that the more centralized a school system is, the worse the achievement of its students, public or private.

In my own state, New Jersey, the board of higher education has ruled that principals of public schools need not have any teaching experience at all, only a degree in management.

Bad decisions are inevitable when decision making is divorced from reality. In this same regard, I quote from a lead letter to *The New York Times* by one Ron Szary, published on November 7, 1988, under the title "Management Mentality Is Killing U.S. Industry." As his first reason for why the United States trailed the Japanese in the superconductor race, Mr. Szary said:

> Unlike the Japanese, American companies are top-heavy with layers of management . . . all of whom want final say on products and production, but have little if any comprehension. . . . In the United States, it would be embarrassing to have the top man in most companies talk to the people on the floor: he doesn't know the product; he doesn't know the process. . . . Our layers of management are calculated to buffer the top levels. On the other hand, it is not unusual to

see top Japanese management in direct, daily contact with workers. Who will outperform whom?

A second consequence of overmanagement is the widespread problem of *demoralization* of the producers. The techniques that help managers expand their power base—the barrage of paperwork which everyone knows is worthless; the deluge of conflicting, often arbitrary, memoranda; the insistence on "accountability" without any standards of reference for performance; the requirement of doing more things than are possible, some of which are in conflict with each other; and other forms of control practiced by management—create numerous double binds in the daily lives of the producer. Some of the techniques resemble scaled-down versions of the practices used to destroy the spirit of prisoners in concentration camps. Primo Levi, writing about his confinement in Auschwitz, said, "The rites to be carried out were infinite and senseless." So too, albeit on a far lower plane of evil, the rules and procedures of overmanaged organizations. I am referring to the lack of information about what is happening and what is planned, the random, fluctuating orders, the removal of sense, purpose, and pride of work from life, and the growing loss of respect for the separateness and privacy of the person.

Not surprisingly, workers in the most seriously overadministered institutions suffer from low morale and despondency. Some of them become physically ill. In such institutions another positive feedback loop is created: excessive administration leads to demoralization, which leads to poor performance, which leads to yet more stringent and pervasive administrative control. For a few producers, the only way to resolve the problem is by dying. For the great majority, the solution is not so severe: some have neurotic ailments, some take early retirement, some thrive, most just do their jobs and endure.

The last consequence of overmanagement that I will discuss is *the decline of science*. This decline, still in its early stages, is not yet apparent to most observers, who perceive the power without noticing the rot that lies beneath the facade.

What is undoing science is the success of its own applications. The public, which pays for much of science, quite reasonably expects it to make practical discoveries of wide usefulness. When this happens in a branch of science, it becomes more powerful and, as a result, more organized. This organization—administration—ultimately stifles the basic science that is the source of invention. For example, undergraduate and even graduate students, for whom the grant-seeking nabobs of science now have no time, are neglected by the very people who ought to be their primary teachers.

To eschew the success brought about by practical discoveries, even if it were possible, hardly makes sense in a crowded world beset with problems that science can address. But for science to initiate its own decline when it makes these powerful discoveries is also not an acceptable state of affairs. There is a tragic paradox in this, and it is not a new one.

In his novel *The French Lieutenant's Woman*, John Fowles wrote:

> We can trace the Victorian gentleman's best qualities back to the parfit knights and *preux chevaliers* of the Middle Ages; and trace them forward into the modern gentleman, that breed we call scientists, since that is where the river undoubtedly has run. In other words, every culture, however undemocratic, or however egalitarian, needs a kind of self-questioning, ethical elite, and one that is bound by certain rules of conduct, some of which may be very unethical, and so account for the eventual death of the form. . . . The scientist is but one more form; and will be superseded.

It was "trade" that killed the ideal of the Victorian gentleman, and it is the quest for a success defined as power that is killing the modern scientist. But the mere quest for power does not kill: in this case, the actual instrument of death is the administrative process created to handle all that power.

To see how the process works, take the case of the Brogdale Experimental Horticultural Station in Kent. Brogdale has the world's best living collection of varieties of apple, not to mention pear, plum, cherry and bush fruits. It is a vital repository of the genes that will be needed by genetic engineers and other plant breeders to build hardiness and resistance to diseases and pests into our next generation of fruit trees. But, in the name of efficiency, administrators of the British Ministry of Agriculture, Fisheries and Food decided to close Brogdale by April of 1990. They promised to keep the collections alive. (In the United States we already have seen, at the National Seed Storage Laboratory in Fort Collins, Colorado, what happens when administrators promise to take care of our precious heritage of grains and vegetables. And as in Britain, the United States government decided to abandon by 1993 one-half of the apple varieties in its extensive collections, to save money.)

Fortunately, in the case of Brogdale a private foundation stepped into the breach, took Brogdale off the government's hands, and set up a Brogdale Trust, which hopes to maintain the collection, although the research program will be reduced. (It remains to be seen how many of the abandoned American varieties will be saved by private organizations.) In addition to abandoning Brogdale, the MAFF also decided to close the Rosewarne Station in Cornwall, which *New Scientist* described as a "unique site for research into early winter vegetables," and was considering closing the vegetable gene bank at Wellesbourne, in Warwickshire.

One of the curious features of today's management, despite its fetish for formal, written goals and objectives, plans, and mission statements, is its inability to comprehend and support any but the most immediate and short-term benefits. When management is brought in to control and enhance the flow of power emanating from modern science, it should not surprise anyone if the goals of administration are substituted for the societal goals that science once appeared to accept. Paradoxically, it is genetic engineering that is helping to power the growth of agricultural administration, yet the same administration, applying its own criteria of value, can jettison Brogdale and other sources of the genes that genetic engineers need for their work.

Overmanagement is also one of the main reasons why science is pricing itself out of existence, and not just the "big science" described by *The New York Times* in a series of articles and letters beginning in the fall of 1990. Projects with high operating costs are favored by administrators because the administration receives a percentage of those operating costs. Research that is spare and economical is often suspect—because it costs little money it brings in little money. Such research, by administrative definition, is not "world class."

In justification of this expensive policy, we hear self-serving arguments that scientific research creates more societal wealth than it consumes. This claim is hard to test, which is perhaps why it is so freely made. In one of the few good studies that bear on the argument, C. West Churchman, the father of systems analysis, demonstrated that the money spent on the Apollo moon program could not be justified by cost-benefit analysis. This study was commissioned by NASA. There is no proof of the assertion that societal benefits are proportional to scientific spending, and no evident reason why expensive research should always prove more beneficial than inexpensive research; often the opposite is true. Are we coming up with more discoveries of lasting value to the public than we did ten years ago? I doubt it. But management, caught up in its own expansion and bewitched by the good news that it chooses to hear, cannot rein in before the crash. Meanwhile, preexisting research and all teaching are in disarray because of the chaos caused by impoverishment and perpetual reorganization, which are hallmarks of modern management.

As resources dwindle and debt mounts, only the most stubbornly unobservant can expect the scientific gravy train to keep on running much longer. Nevertheless, in the coming years we will see a stream of theories and inventions that are supposed to create wealth and power out of inexhaustible commodities such as seawater. Nuclear power was an early example, now curdled and gone sour. Superconductivity and cold fusion are recent illustrations. All will promise much for very little, all would maintain the power and growth of the scientific/technical/managerial system that created them. They are a form of magical thinking bred by a mixture of greed and desperation, and grafted onto real physical, chemical, and biological processes.

Controlling Management

Can we resolve the crisis of overmanagement? True, management, like anything aspiring to perpetual growth, will eventually bring itself under control by running out of resources. It is self-destructing now, as you read these words. The process is being hastened by the tendency of management to cripple the producers, the people who provide the wealth in the first place. Because it incorporates so many positive feedback loops, modern management is inherently unstable.

But the breakdown of the bureaucracy may take more time than we care to wait, and its loss may be small comfort if the world is left a bankrupt ruin. In the past, major attempts to curb the bureaucracy— such as the Carter administration's Reorganization Project—have failed. Can we do anything to shrink management to its proper size before it is too late?

To curb managerial excess, the first thing that comes to mind is shutting off the money tap, or at least slowing the flow. Bureaucratic growth costs a great deal of money. But because the administrators control institutional budgets, it is hard to reduce their funding. I know of only a few cases where this has been done. For example, the Internal Revenue Service effectively limits the percentage of income that a nonprofit foundation can spend on the administration of its grants. The limits are imposed from without. Analogously, an informed Congress might force governmental agencies such as the National Science Foundation to standardize and sharply reduce the overhead given to institutions as part of research grants. Ten percent would be a reasonable figure. If adopted by the major granting agencies such a policy would do much to restore research and teaching in all subjects because managers would no longer be preoccupied with the pursuit of huge grants for their overhead. Nor would there be so many managers. An added dividend is that without extortionate overheads to pay (some exceed 100 percent of the actual grant), granting agencies would have far more money to spend on the research itself.

Besides reducing the money supply for management, other methods of controlling the controllers suggest themselves. Positive feedback loops can be eliminated if producers can participate, at some level, in the hiring and firing of administrators. Even more effective would be the breaking down of the work barriers between bureaucrats and producers through rotation of jobs. In the federal government, this sort of system was pioneered by the Geologic Division of the U.S. Geological Survey. In the USGS, after a few years in management, geologists are often rotated back again into the field as technical scientists. Abuses disappear when management ceases to be a career apart from the work itself.

Secretiveness of managers about their actual work and especially about the details of their administrative budgets is a key element of managerial control. This secrecy is potentially one of the most vulnerable parts of the bureaucratic hegemony. Training a spotlight on administration can be surprisingly effective, as the Columbia University protestors discovered in 1968 when they published their pamphlet, "Who Rules Columbia?" which traced the power structure of the university's administration. This spotlighting effectively reverses the usual technique of administrative dominance, as the producers obtain information about the managers, not vice versa.

Of course such spotlight strategies are not applicable to all bureaucracies, and there is always the problem of who will bell the cat. In many cases of managerial explosion, relief will have to come from outside. This relief effort has not yet been organized, although the force of angry producers is considerable. Both Presidents Carter and Reagan were elected promising to curb governmental bureaucracy, and both ended up by making it worse. These is still no serious strategy for this task, no group of activists uniquely dedicated to its completion. One problem is that to see overmanagement for the danger that it poses requires a revolutionary change in our way of looking at the world: the existing political dichotomies such as liberal versus conservative and labor versus capital do not serve the purpose. A world is waiting to be explored here: strategies for limiting managerial growth have hardly been thought about; the critical relationship between the explosion of management and other modern phenomena such as technology, overpopulation, and urbanization has not yet been examined.

I have no doubt that some kind of revolution against overmanagement will come, but I have no idea who the revolutionaries might be. (I know who they will not be, however: they will not be antiregulatory conservatives, whose motivation is simply to preserve corporate power in the face of governmental limitations and who would be opposed to any real effort to reduce the managerial presence in our society.) No word or phrase exists yet to describe the coming revolution. Perhaps there will not be a word, nor any active revolutionaries—great changes can come quietly, piecemeal, and be visible only in retrospect. Will the managers just disappear, one by one as the money runs out, and not be replaced? If so, we can be sure that many producers will be sacrificed first. Much lingering damage will be done, much chaos will result. To survive with the many good features of our society intact and working well, we ought to solve the problem of overmanagement before it solves itself. But as long as we continue to be a power-worshipping society dominated by the myth of total control, we doom ourselves to some excess of administration and all the misery that this entails. We cannot at the same time worship the idea of control and reject the controllers. Our job, in the words of Lewis Mumford, is to find a way to convert a ''power economy to a life economy.''

Forgetting

On the nineteenth of April 1989, one of the huge gun turrets on the battleship *Iowa* blew up, killing the sailors who were manning it. Debate about the responsibility for the explosion continued long afterwards, but lost in the emotion of the tragedy was a curious aspect of the story. According to the commander of the ship, interviewed after the event, it was expected to be difficult, if not impossible, to fix the damaged gun turret. The *Iowa* is of World War II vintage, and he feared that the materials and technological know-how to repair its gigantic guns might not exist anymore.

A similar problem arose about ten years ago when church officials decided it was time to resume construction of New York's vast Cathedral of St. John the Divine, after a lapse of decades. It turned out that a few old men in England were the only stonemasons left in the world who knew how to work the giant blocks from which a cathedral is built. If they hadn't been able to train young apprentices, there would have been no choice but to abandon the project in a few years.

I think that our concept of progress prevents us from being aware that skills and knowledge can vanish from the world. Most of us probably imagine knowledge to be cumulative: each advance is built on prior discoveries, block piled upon block in an ever-growing edifice. We don't think of the blocks underneath as crumbling away or, worse yet, simply vanishing. Our world view does not prepare us for that, although it is hardly a new phenomenon in the world.

There are many disappearing varieties of knowledge besides the making of guns and cathedrals. Boat designing and building, using a lathe creatively, knowing how to garden without chemicals, having the skill to sharpen a pair of scissors—these and ten thousand other kinds of transmitted knowledge are being lost at a terrifying rate. This grand forgetting could be, and no doubt will be, the subject of entire books. But as a teacher I will write about the losses that are closest to me and my work.

Above all, the loss of knowledge and skills is a serious, even overwhelming, problem in our universities—in the humanities, the social sciences, and the sciences—and no store of learning is in greater danger of disappearing than our long-accumulated knowledge of the natural world. The problem is so acute that I do not hesitate to call it the next environmental crisis, although it will never rival for press coverage the hole in the ozone layer or global warming. We are on the verge of losing our ability to tell one plant or animal from another and of forgetting how the known species interact among themselves and with their environments.

The process is gradual, and it is affecting the more prestigious, research-oriented schools first. What is happening is that certain subjects no longer have anyone to teach them, or are taught on a piecemeal basis by people from the periphery of the university or outside it altogether. "Classification of Higher Plants," "Marine Invertebrates," "Ornithology," "Mammalogy," "Cryptogams" (ferns and mosses), "Biogeography," "Comparative Physiology," "Entomology"—you may find some of them in the catalog, but too often with the notation alongside, "Not offered in 1992–93." The following year, and the year after, they will still not be offered.

The features that distinguish lizards from snakes from crocodilians from turtles from tuataras are not any less accepted or valid than they were thirty years ago, nor are they any easier to learn on your own from books, without hands-on laboratory instruction. But try getting someone in any of the top-ranked biology departments to teach such a lab. In one Ivy League university the school of forestry uses retired faculty to teach some of its basic forestry courses; this same university has trouble staffing a general ecology course from the faculty of its biology department. As I write this, there is a large land-grant university that has no limnologist (a person who studies the biology of lakes and rivers) on its regular faculty of biological sciences, and only one plant taxonomist, retired, on its main science campuses.

New students who are attracted to the study of whole plants and animals still exist, but they find themselves in a very hostile learning environment for their kind of biology. Not surprisingly, their numbers are dwindling. It is these students who, after getting their masters and doctoral degrees, ought to be going out to teach their subjects in the nation's colleges and universities, to be taking over as older professors retire. But the chain has been interrupted: there are many fewer new graduates with this knowledge, and these few will have trouble finding positions unless they are willing to switch to a more "modern" field or to use high-tech methodologies in their research, whether these methodologies are appropriate or not. In any event, the result is the same—reservoirs that are not replenished soon run dry.

To prove that I am not crying wolf, I want to tell a true story. One morning at eight o'clock, my phone rang. It was a former student of mine who was now a research endocrinologist at a major teaching hospital in Houston. She had an odd question: At what point in animal evolution was the porphyrin molecule (such as hemoglobin) first adopted for use specifically as an oxygen carrier? It was an essential piece of information for medical research that she was planning. If I didn't know the answer (I didn't), who did?

I racked my brains to think of a contemporary biochemist or university department that could provide this answer. Nothing. All I could come up with was a book, I thought by somebody named F. A. Baldwin, that I had read when I was a student. She thanked me politely and said goodbye.

Later I went down to the basement and found the book in a box. It was *An Introduction to Comparative Biochemistry,* by Ernest (not F. A.) Baldwin, Cambridge University Press, 1964, fourth edition. The flyleaf, I noted ruefully, indicated that this hardcover text had set me back $2.75, new. Much of the information my former student had wanted was in there, brilliantly written. A phone call to the Cambridge University Press told me that Baldwin had gone out of print in 1980.

By coincidence, I was scheduled to lecture that afternoon to a group of biochemistry professors and graduate students. So I asked them the question I had been asked earlier. "I'm not a biochemist," I said after describing the phone call. "Tell me who is working on this sort of thing these days." They looked at one another and laughed. Nobody does comparative biochemistry anymore, they answered; at least they didn't know of anybody. There probably was nothing much more recent than Baldwin. As for the graduate students, they had never even heard of comparative biochemistry.

Gone! Not outdated. Not superseded. Not scientifically or politically controversial. Not even merely frivolous. A whole continent of important human knowledge gone, like Atlantis beneath the waves. True, we still have Ernest Baldwin's book, but this kind of knowledge needs trained, experienced people to keep it alive and to hand it on to the next generation.

At nearly all of today's research colleges and universities the teaching is being done by three kinds of "temporaries": graduate students; nontenure-track researchers and scholars—mostly women—who work full-time hours for part-time pay and reduced benefits; and an assortment of experts from outside the university who free-lance courses a semester at a time. What they have in common is that they are skilled workers working for substandard wages with no job security. A few don't mind—those who are retired and teaching an occasional course because they want to keep active. But for most it is a matter of survival, and they tend to feel exploited and to become angry, depressed, or a mixture of the two. Some of these teachers manage to be conscientious, inspiring, and creative, but few are around for very long. Teaching, more than other professions, needs continuity.

Despite the starvation of teaching, universities are receiving and spending money as never before. Where is it going? The answer varies from school to school—at one it will be computer science, at a second genetic engineering, at a third high-energy physics—but in all cases the money is going to hire "world-class scholars" at world-class salaries, and to set them up in business. At my university, as at many others, world-class scholars have become a kind of academic consumer item, like fancy computer systems; our administrators refer to them as WCSls, or wixels. They are purchased on the open market. One wixel can cost tens of millions of dollars by the time the university is finished providing the building, space-age equipment, and numerous support personnel that the wixel has been promised. Wixels don't have time to teach, except for occasional cameo performances; they don't even have time for graduate students.

Eventually, every asset that the university can lay hands on is hocked to pay for these wixels. Teaching budgets are slashed, teaching laboratories are converted into research space, and the salaries of professors who were foolish enough to teach or whose research is not in one of the glamorous areas are seized when these professors retire or, if untenured, inevitably fail to gain promotion. Soon, the only way the university can afford to keep its teaching program afloat is to hire a flock of temporaries. Not only are they cheap, but if they complain they can be fired.

Common sense tells us that starving the roots of a fruit tree to promote the bloom of a few showy and mostly sterile flowers is irresponsibly short-sighted. What has driven higher education to commit this folly? The ultimate reason why we forget knowledge that is important to us is the corruption and loss of our societal values, particularly those values associated with peace and permanence. Values remind us of the importance of certain things in our lives, including special varieties of knowledge. Without values our attention to life wanders—is it so strange, then, that we lose those things that we have forgotten we need? Yet if the ultimate cause of forgetting in our universities is a problem of values, the immediate, practical cause, as you have surely guessed, is university administration, with its uncontrolled growth and insatiable demand for funds.

Before the Second World War, universities were run by a rather small cadre of scholars turned administrators, usually distinguished professors who had reached a point in their careers where pomp and ceremony were more appealing than the library or the laboratory. This was harmless—even useful. Every university needs a royal family to get money and charm the public. But after the war, things began to change. The Managerial Revolution was upon us, university administration became a career in itself (especially for those whose academic work wasn't going anywhere), and administrators proliferated like weeds in a garden. Worse yet, they assumed the salaries and office luxuries of the corporate executives they were

imitating: the royal family became a royal mob, losing both its no-
bility and its utility in the process.

The new administration had new priorities, most of them dictated
by an overwhelming need to find money. But taxonomy and other
classical fields of biology get comparatively few and insubstantial
grants. The overhead these grants provide is minimal. The classical
fields simply do not support enough administrators to make the fields
worth keeping. For this reason, whole disciplines are adminis-
tratively branded "not world class" or "unproductive," and are
dumped, like trash, from the university canon. The word gets
around; administrators listen to each other, if to nobody else. School
after school jettisons the stigmatized disciplines rather than have its
administrators labelled as not progressive.

The money chase is a nerve-wracking business. Grants and patents, with their lucrative, easily laundered overheads, are not
"hard" budget items. A bloated administration requires more and
more of these scantily monitored funds to support its growth, but
grants and patents are undependable. Inevitably, universities begin to
bid against one another to attract those scientists (the wixels) who
have the best records of getting large grants.

University administrators now find themselves on a treadmill that
they cannot get off. Wixels almost always cost more than they bring
in, if *all* the costs, including extra administration, maintenance of
new buildings, etc., are taken into account. This is true even of the
good wixels, let alone the duds. Fortunes are spent to gain fortunes
that never materialize. Student tuitions are raised and raised, "unproductive" departments are closed, budgets (except the wixels') are
pared. Many universities, despite massive endowments and cash
flows, are now stretched perilously thin. The system is spiraling out
of control.

Here is a very ominous positive feedback loop, like the other
positive feedbacks that accompany overmanagement. The fewer research programs and courses there are in a subject, the fewer people
there will be to teach the next generation of students. Skilled knowledge is transmitted by people, not by books alone. Make no mistake,
I am not talking about the preservation of trivia, but the safe transmission of vital existing knowledge. An example: agriculture depends on soil. Soil fertility, as Darwin knew, depends heavily on

earthworms; many different species of earthworms play different roles in soils. In North America, a quiet but potentially momentous battle is being waged as European and Asian earthworms displace native species. The invaders seem to be quite different ecologically from the local worms, so consequences can be expected, even if we cannot predict them. Yet at the time of this writing, there is just one actively working scientist who is familiar with the taxonomy of the earthworms of North America. He is at a small private university in Iowa. Another earthworm taxonomist works at a university in Puerto Rico, but she was only recently trained in Spain. A third earthworm taxonomist, trained by his mother, has been working for a post office in Oregon. The fourth, and last, person in North America north of Mexico who has expert knowledge of earthworm taxonomy is presently earning a living as a police lawyer in New Brunswick, Canada. There are no more graduate students studying earthworm taxonomy in the United States and Canada. Fifty years ago, at least five American scientists, plus their students, were at work in this field. Nor is the situation different in other parts of the world: Australia, long noted for earthworm research, now has none; the British Museum has ended its earthworm taxonomy, and so on.

The example of earthworms is not atypical. The more advances we make, the more we forget. What use is our expensive technology in a sea of ignorance? It is time to end the Managerial University while there are still people left in it who know how to pass on useful information from one generation to the next.

Nor is the problem confined to the sciences. In the social sciences and humanities it is not necessarily big grants that have derailed the transmission of knowledge, but a similar influence. These divisions of the university have their wixels, too—showy, fashionable, academic superstars who thrive on publicity and can attract the attention of big donors, thus reflecting glory on the administrators who hired them. Those who suffer more directly than I can expose these follies.

As in other kinds of overmanagement, the way to start braking the administrative juggernaut is to reduce the flow of money to administrators—soon. There is, for instance, no reason why unspecified grant overheads should exceed eight or ten percent. In the case of heavily endowed schools, wealthy alumni should also put an end to knee-jerk giving, especially contributions for new buildings. In the modern university, money is increasingly proving to be a corrosive substance.

Turning off the money supply is not enough, however. Teachers need the courage and skill to publicize the importance and the plight of the knowledge that is vanishing. An informed university community may then be roused to demand cuts in administration (starting with higher administration), less secrecy, greater faculty and student influence, a moratorium on construction of "high-tech" facilities and on increases in tuition, a higher priority for teaching, greater communication and interdependence between the university and its surrounding community, and support for a diversity of low-cost research projects which can function without multimillion-dollar grants and which may not generate profitable patents. And none of this will happen unless we constantly inform people and explain our efforts in the light of the values we seek to recover.

But if no effective change takes place, what then? Then we can expect the managerial ethic to continue to prevail and teaching to become vestigial as the existing university structure falls further into disarray. True, a new kind of university may emerge, perhaps already is emerging. It will have some positive features. But whatever its virtues, it will not be capable of preserving and transmitting our assembled knowledge of the natural world. I fear for us when there is no one left in our places of learning who can tell one moth from another, no one who knows the habits of hornbills, no one to puzzle over the diversity of hawthorns, no one even to know that this knowledge was needed and is gone.

State of the Art

The trouble began one day about three or four years ago when I was sitting in my former office in Rutgers's Old Blake Hall. Old Blake was built more than a half-century ago by New Jersey's fruit growers to house the ag school's pomologists—scientists who study how to grow fruit. The building was named after Professor Maurice A. Blake, an early Rutgers pomologist who knew all there was to know about peach trees. But universities have short memories when it comes to gifts from people who are now dead, and when I was there, Old Blake housed Forestry and Wildlife, the group with which I had cast my lot.

Before telling about what happened, I should describe my ground-floor office in Old Blake. It measured about twelve by fifteen feet and had a linoleum floor, one tall, double-hung window with an air conditioner mounted at the top, a steam radiator under the window, and—sorry to make you keep looking up and down—a very high, vaulted plaster ceiling with pipes coming out of it leading to and from the men's room upstairs. The walls were immensely thick, holding the coolness of spring well into summer. The room was crammed with two desks, three chairs, a typing table, four filing cabinets, a closet in one corner which left the remaining open space L-shaped, two low bookcases, and, bolted to the wall, a nine- or ten-foot bookcase whose upper shelves were reached by an attached, magnificent, varnished oak ladder on wheels. My children loved to climb it.

It had been made to order for me by the Putnam Rolling Ladder Company of New York City.

The window (pay special attention now) could be opened from the bottom about four inches. A stop prevented it from opening farther and thus kept out the sort of folks who like to enter locked buildings through windows on weekend nights. Wasps could get in, however, although they found it very difficult to get out again. A pair of mourning doves nested outside, under the air conditioner; I could hear but not see them. Occasionally, bits of nesting material were shoved through cracks in the insulation around the air conditioner and fell on the plants that lined the radiator cover beneath. One of the plants, Herman, a philodendron associated in its pot with a climbing fern, traced its origin back to the greenhouse at Barnard College, where I had previously taught. My view out the window traveled eastward, past some bushes, across a narrow, untrafficked road, past a row of red oaks, and on across the great central lawn of the college, which was dotted with clumps of exotic trees such as cedars of lebanon. The lawn sloped northward down to the college pond, Passion Puddle.

As I was sitting in that office, my workplace for fifteen years, one of my colleagues came in and told me the news: Forestry and Wildlife was being merged with the Department of Environmental Resources, and would be moved out of Old Blake into the new Natural Resources Building that was being planned. Marine and Coastal Sciences, a highly favored new unit at Rutgers, was coming into Old Blake.

"The university is spending more than nine million dollars on the Natural Resources Building," my informant said. "It's going to be a showpiece, a state-of-the-art facility."

My blood chilled. "State of the art," I quavered. "Will the windows open?"

Not hiding his disgust at my cynical and selfish attitude, he told me that selected representatives of the faculty were even now helping the architects and engineers plan the new building. Just then the toilet upstairs flushed, sending a rich gurgle through the pipes overhead. Turning hastily to leave, he commented that I ought to be delighted to leave Old Blake.

The planning sessions occurred frequently. I was informed that several faculty representatives had requested openable windows and that this was a "consensus position." After all, it wouldn't do to have the Natural Resources Building hermetically sealed.

Eventually, ground was broken and construction began. I slogged through the mud and found the foreman in his trailer, surrounded by plans. He was very cooperative. No, the windows, of tinted glass, would not open. Yes, they had already been ordered. It was too late to change the plans. I disagreed.

An emergency meeting was called about the windows. Three or four of us from the faculty were present, along with representatives of the architects and the firm that designed the climate control system, and a fellow from the Rutgers office that makes sure that all buildings are planned and built properly. They were authoritative. They had plans and diagrams. What they told us was this: First, the "HVAC" climate control system would not work with openable windows. Second, maybe the system would work with openable windows but it would be harder to balance. Third, replacing the windows that had been ordered would cost sixty thousand dollars. Fourth, construction might be delayed a few weeks. So it was up to the dean to decide what to do.

Several months later, we moved into the new building. I left Herman and my other plants behind because not enough light to support photosynthesis came through the tinted glass of my north-facing, unopenable windows. It was winter and my view across the parking lot was bleak. Some of the offices, including mine, were very cold. This was because the air that came through the ceiling vents was chilled, winter and summer. Small radiators worked unsuccessfully to overcome the chill from the vents and the chill from outside. If the thermostats were jimmied open and set to a higher temperature, the flow of chilled air was reduced, but a better result was obtained by stuffing up the vents with tissues, if you could stand the howling of the pent-up air rushing through the narrow opening that remained. Then the room would get warmer but also stuffier as the oxygen gave out. Unfortunately, the windows were sealed too tightly to allow any reviving drafts, although as one of my friends remarked, pointing to a heavy chair, the windows were in fact openable—once.

In summer, things got much worse, especially on the sunny side of the building. Engineers and maintenance people came and went in a frothing stream, responding to a frantic series of phone calls from hot and suffocated inmates. The maintenance staff helped with terminology: "We're not allowed to call it fresh air," said one harassed mechanic. "We have to call it outdoor air."

The chief designer of the climate control system appeared—a little man with big mustaches, surrounded by soothing subordinates. "They open the blinds," he protested angrily. "This system was designed to work with the blinds closed." His subordinates nodded in unison to show their outrage at thoughtless people who open blinds when the sun is shining.

Then the Associate Director of Operations from the Rutgers Department of Radiation and Environmental Health and Safety issued a report. The "numerous complaints" from the Natural Resources Building were "focused on . . . poor temperature control, lack of air movement, excessive air from some diffusers, [and] miscellaneous health problems (headaches, eye irritation, fatigue)." His inspectors found that the dampers were closed on the outdoor-air intake vent, so no outdoor air at all was being circulated in the building. This was perhaps just as well because they also found that the air intake port on the roof was located too close to the building's exhaust. Moreover, they noted that when the temperature in a room matched the thermostat setting, airflow to the room was automatically stopped. From these findings came a few helpful suggestions. My favorite was the plan to install heating coils in the air ducts to heat the chilled air in winter. The idea of retrofitting the building with one openable window in each office was sternly dismissed because of the state's budgetary crisis.

Why does this sort of Rube Goldberg monstrosity get built by a modern university? How can a dysfunctional building cut off from outside air, light, and life be called the Natural Resources Building? The answers are not hard to find. It happens because of the fragmentation and bureaucratization of responsibility and the proliferation of experts, ensuring that no planner will see the entire picture. It happens because of modern, top-down management, in which isolated administrators playing their cards close to the vest make decisions for other people who will have to live with the consequences. And it happens because of the foolish infatuation of decision-makers with "state-of-the-art" systems designed with utter disregard of the needs of their users.

In what was the Soviet Union, an outraged populace fed up with decades of waste, the destruction of human cultures, and the misuse of human resources has started to throw out some of the arrogant central planners and managers. It is an immense task. In America, we are also faced with a glut of managers, substituting their dissociated, power-oriented fiat for the experience and judgment of the people they manage. Our universities, like our government, our industry, and our other institutions, are sorely beset with the blighting curse of imperial management, but no revolution is any closer than the distant horizon. As in the case of loss of knowledge, we may be stuck with "state-of-the-art" design until the money for it gives out.

Meanwhile, please come and visit me in the Natural Resources Building. You can tell it by the two rows of unfinished columns flanking the walkway to the front entrance. I'm told that the columns will some day support a decorative canopy which will shield people from some of the rain until they get near the door. You can tell it by the huge, main stairwell which looks like a yuppie version of the central gallery of a maximum security prison. This is the stairwell that was originally built with each step pitched downwards. It sure was noisy when they drilled out and recemented those stair tiles. You can tell it by the tiny standard offices and the few odd-shaped larger ones, cleverly built to save space and the taxpayers' money. You can tell it by the fancy hallway light fixtures with the imported incandescent bulbs that cost fifty dollars apiece to replace. You can tell it by the corridors that meet at thirty- and sixty-degree angles, causing a novel kind of disorientation. These corridors enclose an interior space that houses a classroom without windows and secretaries' nooks without windows, privacy, space, security, or good lighting. If you're trying to find that classroom, the only one in the building, I must tell you that the sign over the door reads "Research Laboratory." Undergraduates shouldn't get the idea that they are welcome in a state-of-the-art research facility. And you ought not miss the graduate student room, with its outside wall made of little glass blocks that distort and multiply the tantalizing images of trees and people on the other side, in the outdoor air. This window wall helped the architect win an award for design.

If you get to my office, you will see that I really shouldn't complain. It's larger than most offices in the building, and the ceiling is conveniently low, so I have no trouble reaching the air vent when I have to make an adjustment. I even have room for a ten-gallon fish tank—if I turn the filter up, the bubbling masks the sounds of people breathing loudly and other noises on the far side of the wall that separates me from the corridor. Some day I will get some fish for the tank. Best of all, I am not far from the main entrance: every time somebody comes in, a thin but refreshing draught seeps in under my door.

I look forward to your visit, and to the opportunity of showing you around our award-winning building. But call me before you come; I may be working at home.

The Lesson of
the Tower

The farthest north of the state-of-the-art mentality is the drive for the "conquest" of space. Our exploration of space has been accompanied by all the familiar signs of state-of-the-art technology: it is fantastically expensive, much of it is incomprehensible except to experts, it requires a massive amount of organization and management, its true agenda is often hidden, and it cannot be controlled, or even significantly influenced, by the ordinary people who pay for it and must live with its consequences. Space, a totally user-unfriendly environment, is indeed the appropriate subject for a technology that has elevated user-unfriendliness almost to an art form.

The Bible does not record how the inhabitants of Babel felt when they were forced to leave off building their tower and city. The account is terse: the language of Nimrod's people was confounded and they were scattered "abroad upon the face of all the earth." We are left to imagine the passions and the rationalizations, the confusion, denial, anger, despair, and above all the fear that must have accompanied the ruin of so great an enterprise and the overthrow of the accustomed purposes of daily living. That we can well imagine these feelings despite the vast gulf of time and culture that separates us from the people of Babel is because of the fundamental similarity of our predicaments. But we have the advantage if we will use it. We are already familiar with the story, and this may help us modify the outcome.

Since *Challenger* and Chernobyl it is no longer reasonable to doubt that we are entering a new phase of human civilization. The brief but compelling period of overwhelming faith in the promise and power of technology is drawing to a close, to be replaced by an indefinite time of retrenchment, reckoning, and pervasive uncertainty. At best, we will be sweeping up the debris of the age of unbridled technology for decades, perhaps for a longer period than the age itself endured.

Among the many parallels between these times and the days of Babel, the most obvious is the idea of challenging the heavens. The conquest of space has been the apotheosis of a massive technology funded directly and indirectly by citizens and used to further the aims of the state. Considering the duration of the space age for more than a third of a century, its primacy in national and international affairs, and the way it has affected our lives, surprisingly little intelligent thought has been devoted to it. But that is now changing, and one of the first signs of this change was a remarkable and eye-opening book, . . . *The Heavens and the Earth,* by Walter A. McDougall, a history professor at the University of California at Berkeley. Written before *Challenger,* McDougall's book is nevertheless where we can begin our search for an understanding of a rocket—and a civilization—that went so tragically off course. Now, with the expiration of the Soviet Union, which began the space race with the launching of the first *sputnik* on October 4, 1957, America, the surviving superpower, had best think hard about the cost and consequences of a race with only one runner. Will we still want to run this race when we will have identified the prizes? The only prizes we know anything about in advance are the whopping bills for the economic and environmental costs of the conquest of space, and these totals are yet only partly tallied. Although McDougall and I differ over the lessons we draw from the history of space exploration, I acknowledge his work as the prime source of information and inspiration for my thoughts about space technology and its effects.

The thrust into space began with a dream which in its origins goes back beyond Daedalus and his ill-fated son. The modern version, inspired first by Jules Verne and later by H. G. Wells, was soon incorporated into the scientific consciousness. It found one of its earliest expressions in the 1883 diary of Konstantin Tsiolkovsky, *Free Space,* and in his paper, published in 1903, on the "Exploration of Cosmic Space with Reactive Devices"—the foundation of the principles of rocketry. From Tsiolkovsky, inventor of the multistage rocket, who died in 1935, the dream followed an erratic but unbroken course of development that recalls such names as Robert Goddard, Yuri Kondratyuk, Sergei Korolev, and Werner von Braun. In a quotation from Tsiolkovsky (1911) that McDougall uses to introduce the first part of his book, we are exposed to one theme, certainly the original theme of the exploration of space:

> If we can even now glimpse the infinite potentialities of man, then who can tell what we might expect in some thousands of years, with deeper understanding and knowledge.
> There is thus no end to the life, education, and improvement of mankind. Man will progress forever. And if this be so, he must surely achieve immortality.

That this achievement of immortality would be brought about in good part by the conquest of space, Tsiolkovsky and the others who shared his dream made absolutely clear.

Tsiolkovsky, Goddard, and the rest never had any real political power, as McDougall demonstrates. Korolev, who gave the Soviets the first ICBM as well as their greatest triumph, *Sputnik,* did much of his early work while a prisoner in the gulag, along with the rest of his technical team. Even after the triumphs at Tyuratam, the remote Soviet rocket base in Kazakhstan—built by slave labor—Korolev was never entirely secure, never in his fifty-eight years free to follow the dream where it would lead. For there was, both in the Soviet Union and in the United States, another dream—"objective" is a better if uglier word. Tsiolkovsky's dream, which received the attention of the public relations specialists and the media, was embodied in the phrase "one small step for man, one giant leap for mankind." The other objective, we might say the real one, was shared by the men in charge, Nikita Khrushchev, John F. Kennedy, Lyndon Johnson, and their successors. This was the objective behind what the Soviets and then everybody called the Scientific-Technical (S-T)

Revolution. Lenin expressed it best, in 1919, in a quotation also cited by McDougall: "We must master the highest technology or be crushed." In other words, the function of a space program is to enchance the political and military standing of the state. It is the irreconcilable difference between the dream of Tsiolkovsky and the purposes of the S-T Revolution that explains what is now happening in space, and a good deal else besides. It even explains, in part, the self-demolition of Lenin's political legacy, Soviet communism.

Biblical commentators have long been fascinated by the tantalizingly brief account in Genesis of Nimrod, the "mighty hunter," ruler of Babel. Tradition has it that he was the first tyrant. Rabbi Samson Raphael Hirsch, a leader of nineteenth-century European Jewish orthodoxy, declared that Nimrod was the prototype of the dictator who uses a pious hypocrisy to increase his oppressive control over the people. Moreover, the obvious connections in the Babel story between tyranny, technology, unbridled pride and self-adulation, and urban centralization of power have not been overlooked, especially by the modern critics. Again the parallel with the history of space exploration is inescapable. On the one hand are the pious statements of the dream: the drive to explore the unknown, the need to visit other worlds and perhaps find other life in the universe, the desire to unite humanity in a common goal of an essentially peaceful nature, the importance of discovering and developing new resources for the benefit of the suffering populations on earth, and finally, the chance to fulfill ourselves and achieve our own unlimited potential. On the other hand we have the real reasons why space technology has been so lavishly funded by the major industrial powers: the desire to develop new methods of international spying, the importance of mastering the new symbols of power in the court of world public opinion, the need to bolster a sagging aviation industry, and the long-term goal of extending military and political hegemony through the use of new weapons systems in space. On the one hand a dream, on the other a grim and intensely practical objective, with little or nothing in common to relate them to each other.

Perhaps it might be argued that this is the way the world works, the way things get done, with a dream up front, in the limelight, and cigar-smoking power brokers making things happen behind the scenes. Reality is neither one nor the other, but an indefinable mixture of the two. This is how cathedrals and constitutions come into being, so the argument runs, and I have no quarrel with the general idea. But regardless of its applicability in some cases, I don't think it has much bearing on the age of space technology, where the dream itself was badly conceived, poorly developed, and inexpressibly shallow, and where the offstage reality has been from the beginning more cynical, more dangerous, more ruthless, and less connected to the dream than most people have thought possible. The dream I will consider later, after I take up the subject of images—first the offstage reality.

Much of McDougall's book is preoccupied, as any history of the space age would have to be, with the unseen motivations of the people and organizations that made it happen, both here and in the Soviet Union. In this account, a genuine, old-fashioned, and very convincing hero does emerge from the pages of . . . *The Heavens and the Earth*—President Dwight Eisenhower, who resisted the S-T Revolution and the call for a massive, federally funded space technology, who was fully aware of its many dangers, especially for U.S. democracy, and who had little choice but to start the space program anyway and to usher in the era of state science and technology that he distrusted and feared.

Eisenhower, as McDougall portrays him, was wholeheartedly committed to the essentially conservative idea of getting the government out of the lives of its citizens. In this sense, Ike's concept of government was the opposite of that of the Soviet Union. His principal weapon was the budget. After the election that put Kennedy and Johnson in the White House—largely on the strength of public fears about *Sputnik* and the menace of Soviet ICBMs that did not yet exist—the lame-duck President Eisenhower vetoed the multibillion-dollar Apollo moon program. McDougall writes:

> Supporters compared lunar exploration to the voyages of Columbus, but Ike insisted that he was "not about to hock his jewels."

Nor was the ex-general worried about falling behind in the weapons race. When questioned about "catching up" in space, he replied:

"I am always a bit amazed about the business of catching up. What
you want is enough. . . . A deterrent has no added power once it
has become completely adequate."

And public relations did not have the same allure for him as for later
presidents. McDougall states:

Flashy public relations, observed Ike [in a *Saturday Evening Post*
article in 1964], persuaded Americans that labels were somehow
solutions: New Frontiers and Wars on Poverty. Such panaceas usually turned out to be new channels to siphon off power to the federal
government. The space program had started with a step-by-step approach and was now blown out of all proportion by hysterical fanfare.
These were the trends, and soon it would be too late. . . . One day
historians would record that "here was where the U.S., like Rome,
went wrong—here at the peak of its power and prosperity when it
forgot those ideals which made it great."

The tragedy for Eisenhower was that he could not permit himself to
heed his own warnings and follow his own intuitions. Not only was
the political force of public opinion after *Sputnik* too great to resist,
but, ironically, he himself had foreclosed any alternatives by deciding that Truman had been spending too much on the military and by
opting instead for high technology nuclear deterrents, which at the
time were cheaper than conventional armies and weapons. Ike had no
choice but to ring up the curtain on the biggest technology performance of all. What followed was NASA, spy satellites, the moon,
Soyuz, MIRVs, commercial satellites, Star Wars, and *Challenger*.
McDougall looks back on the beginnings of it all for America:

An "honest" space program might have been one single, coordinated
effort run by the DoD and pursued in outspoken competition against
an "inferior, flawed Communist rival." This would have been can-
did, but would also have been a mirror image of the Soviet posture.
U.S. space institutions at least reflected the values of free, open,
international inquiry and discovery for the elevation of the human
spirit, even if deeds did not always measure up. The United States
under Eisenhower traveled far on the road to technocracy, but it still
sheltered the memory of goals loftier than those of the power state.

After Eisenhower, the memory soon faded and was replaced by rhetoric. If there were any scruples about technocracy under Kennedy, Johnson, and Nixon, they were not readily apparent to the historian. Stirred as he is by the more visible achievements of the space race, McDougall has no illusions about their cost, especially to the United States. The cost, as he describes it, was more than dollars, although that cost was massive enough. It was the substitution of technicians for soldiers and a defective systems analysis for military judgment. It was the creation of a bloated research and development empire run by men like James Webb of NASA, grandest of the postwar "Big Operators." It was the militarization of a civilian economy. It was the application of a technocratic and frontier mentality to nearly all areas of hitherto unaffected civilian life. It was the creation of an "act" so spectacular that it made other excitements in life seem boring and trivial by comparison and made itself impossible to follow with merely more of the same. It was the restructuring of academic science, indeed of all academic life, and the changing of the nature of the modern university from a house of learning to a place where money is extracted from the practice of technology. It was the growing reliance on think tanks for planning national strategy, and the introduction of terminology such as "flexible nuclear response" and "mutual assured destruction." It was, perhaps, the throwing away of "the last chance to halt the missile race at a primitive level." It was the perfection of the dangerous idea of technology as a magic wand, a cure for all problems. And it was the institutionalization and regularization of the practice of concealing reality through the use of a series of ever more polished and elaborate images.

Above all, the history of the American space effort since Eisenhower is a history of the manipulation of images—images used to conceal the real purposes of the people who were pushing us into space. The preoccupation with images began with Kennedy, who first perfected the substitution of symbols for reality that has increasingly characterized our national life and reached a new stage of development under the actor-president and his imitative successor. McDougall does not dwell on this theme, but as I examined his factual descriptions of the political process that resulted in the embracing of the Apollo moon program by the President and Congress, I recalled the image of the tall, boyish-looking president with his designer hairdo, his beautiful cosmopolitan wife, and the air of Camelot that surrounded them both. Kennedy was a public relations triumph, a master of images, and thoughts about images seem to have dominated a good part of his decision making. When Jerome Wiesner's Ad Hoc Committee for Space advised him that he should dissociate himself from the Mercury program because of the risk of failure or death of an astronaut, he readily complied. McDougall speculates:

> The admonition [in the committee's report] that seemed to affect the new President the most was that concerning Mercury—exploding rockets, dead astronauts, lost races—that is, not that manned spaceflight was wasteful or misguided, but that it might be a public relations failure.

Nevertheless, he soon reversed himself on the issue of manned space flight and, shortly before his assassination, on the decision to make an all-out effort to put men on the moon. The debates within the administration are carefully traced by McDougall and there is only one reasonable conclusion to draw from them: Kennedy, then Johnson and Nixon, supported the race to the moon (as opposed to space efforts that were more low-key) for its public relations value. The international and national prestige that success would bring, and the disgrace that they assumed would come if we failed to compete, left them, they believed, no choice. The implications of this thinking are analyzed by McDougall:

> Throughout the century, Soviet leaders declared that the Western countries might be ahead for the moment but were run by and for monopolists who lived off imperialism foisted on others by their high technology. The USSR, on the other hand, was a people's state . . . and in any case would soon catch up. Now the United States was saying that the Soviet Union might be ahead for the moment but was run as a slave state in which resources could be mobilized to serve militaristic ambitions. The United States, on the other hand, was a free country . . . and in any case would soon catch up. In so saying and doing, American leaders bought the Communist line that

technology was both a symbol of social superiority and the main agency for making one's own preferred image real.

There were as many reasons for the selling of space as there were government officials and agencies who promoted it. For James Webb, the head of NASA in its formative years, Apollo was a way of teaching us how "to organize the complex and the unusual," how to manage the "large aggregations of resources and power" that were, he felt, the key to American progress and survival. For Robert McNamara, Kennedy's secretary of defense, NASA and the space program were a chance to reorganize the military, subdue interservice rivalries, and pump new life into an aviation industry that was undergoing one of its periodic crises of income. Finally, the most obvious reason of all, the American public might not be ready to spend scores of billions of dollars and turn its society inside out to develop vast new military systems during peacetime, especially during *détente*. A State Department draft of 1961 states: "Military space programs, even in research and preliminary planning stage, should be played down publicly." For many in government, space imagery was the means for concealing a primary military purpose. The Soviet space venture was entirely military in nature, an ugly truth that was concealed by simple denial and lying, a deception made easier by the mantle of secrecy shrouding their whole operation. The American program was primarily military in nature, a fact that could only be concealed—in an open society—by hoopla: the manned moon landing, the cooperation with other nations' space efforts, the space shuttle, and the much-touted spin-off benefits of space technology.

Images fade and have to be retouched. Space imagery was no exception. It was impossible to hide all the colonels, majors, and captains who peopled NASA's flights. Moon shots became boring. The few foreign astronauts on U.S. flights hardly constituted meaningful international cooperation in space. The space shuttle, which carried heavy payloads into low orbits particularly suited for military missions but not useful for much else, was hard pressed to find any other scientific *raison d'être*. And, as I have noted in an earlier chapter, NASA's own consultant, C. West Churchman, showed that

Apollo could not be justified on the basis of an economic cost–benefit analysis—military technology produces few civilian, commercial benefits. Who, then, was fooled? Not the Europeans or the Japanese, and certainly not the Russians. McDougall answers the question:

> The only customers for the unrealistic rhetoric, the only ones who may have believed in it, were the American people themselves.

NASA was stuck with imagery that was becoming increasingly unworkable. After Vietnam and Love Canal it was faced with a citizenry that no longer took government at its word. Meanwhile, the Soviets had chosen early on not to try to put a cosmonaut on the moon. By this decision they avoided the scientific-technical dead end of Apollo-style, manned lunar voyages, and escaped the limitations imposed on the American programs by its own public relations success. But America was committed to manned space flight of a particularly showy nature, which more often than not was an albatross around the neck of NASA. It was becoming harder and harder to amuse the American public.

As NASA's budget increased in 1985 to more than seven billion dollars to accommodate the research demands of Star Wars, it trotted out a new gimmick. A teacher would be sent into space, an appropriate gesture in a country that has increasingly confounded education with entertainment. McDougall has nothing to say about *Challenger*—his book was published ten months before it went down. Even so, . . . *The Heavens and the Earth* gives us the understanding of American space images that lets us ask the right questions about the *Challenger* mission.

What are the right questions to ask about *Challenger?* They are certainly not the "how" questions, the questions about the mechanism of the explosion that have preoccupied the press since the day of the accident. The right questions have nothing directly to do with O rings, booster rockets, cold temperatures, or even mismanagement, although NASA was happy to confine the inquiry to those issues. Actually, there is only one right question: Why was a schoolteacher sent up into space, and what was the purpose of what she was supposed to be doing there? According to *Life* magazine, two months before the event, her mission was "to demystify space travel." One way she would do this was by shaking up a plastic bag containing raisins and marshmallows, in "an experiment about gravity" which she was to perform on television from space. More gimmicks. Would she, I wonder, also have shown the students back on earth the experiments her military fellow passengers were doing in *Challenger?* Were they also to have been working with raisins and marshmallows?

The question NASA could not afford to have asked was, why send a teacher into space? Does a teacher teach better from a TV tube than in face-to-face contact with students? Is there something special that students can learn about from space? That one gets sick to one's stomach up there? They know that. That there is a magnificent view? They know that, too. That the human spirit is very special? They could learn that better, if they wished, closer at hand, by studying the writings of survivors of the Holocaust, or, on a more mundane level, by simply looking around them at the countless examples of the bravery and humor of ordinary people coping with the stress of daily life in space-age America.

Why then did they send a teacher up there—a brave if somewhat misled person who ought still to be alive and teaching here on earth? The answer can only be more image-polishing. People who watch, on television, a personable and dynamic teacher shaking raisins and marshmallows together in a bag while traveling in a spaceship are not likely to ask questions such as: Why do we have a manned space shuttle in the first place, and what are the costs that all of us must bear?

The space imagery that still prevails in the United States is only a caricature of the dream of Tsiolkovsky, a little like the Marx brothers singing *Il Trovatore* (except that the Marx brothers were never inept). Unfortunately, the preoccupation of McDougall and other thoughtful critics with the imagery has served to divert criticism from the dream itself. McDougall never comes to grips with this in his book; busy with the imagery, he appears to accept the dream at face value, even to be dazzled by it. To be fair to him, it would have been a terrible book if it had been written by a person who couldn't thrill to the idea of space travel. Moreover, this is a pervasive and appealing fantasy for almost anyone who reads or watches television. Who has not been at some point a vicarious traveler in the glorious space voyages of H. G. Wells, of C. S. Lewis, of Isaac Asimov, of Robert Heinlein, of Ursula Le Guin, or of the starship *Enterprise?* But the dream of Tsiolkovsky was never intended to be fiction, not even science fiction.

The dream of the mastery of space has become, if it wasn't before, inappropriate, self-destructive, nihilistic, and in the real sense of the term, anti-life. Dreams are a fundamental and necessary part of human nature, but this does not give us license to pursue them anywhere at any cost. Dreaming is a noble trait; but not all dreams are noble, especially in their execution. It is time we realized that aspiring to go to the top of Mount Everest or to the bottom of the ocean, which are on this planet, has a different meaning and far different consequences than trying to get to the moon and Mars.

The fact that the dream of space travel can give pleasure is irrelevant; the point is that it was never very well thought out. For those artists who have used space travel as a way of conveying some unrelated message or helping to explore some unrelated idea or adventure—Lewis and Le Guin are two such artists—this doesn't matter. For example, Hugh Lofting in his extraordinary Dr. Doolittle series has the doctor and his companions take a trip to the moon on the back of a giant moth. The exotic creatures and vegetation that he finds there are reminiscent of the exotic creatures and vegetation that he encounters on his terrestrial voyages. Why did Lofting send Dr. Doolittle to the moon? Probably because he had run out of devices such as floating islands, lost cities, and remote continents here on earth. But in reality, the good doctor would have had a terrible time on the moon as it is, the moon that Neil Armstrong stepped on.

Space fiction has no chance of success unless it is filled with replicas, facsimiles, or at least analogs of life as we know it here on our own planet. The two "constructors" who are the comic heroes of Stanislaw Lem's *Cyberiad* are mechanical devices, advanced cybernetic mechanisms, but anyone who didn't know this and picked a story from the book at random would probably assume they were humans. This is no problem, because Lem is at heart a satirist of the human condition. Space fiction is filled with humanoids, pseudotrees, freakish animals, and mythical microbes, because that is the sole way to make it interesting. The trouble only comes for those who take this sort of thing literally.

The overwhelming reality of space is that it is, for us, for the foreseeable future and perhaps forever, utterly lifeless, bleak, and empty. Nor do we have the ability to make it otherwise, except on paper or on the television screen. We live in the world where we arose, completely suited by God, evolution, or both, to its conditions. Unless we abuse it terribly, this world keeps us alive even if we forget about it or ignore it. When our created systems malfunction, as they always do sooner or later, the earth is still there to hold us and keep us while we tinker with our broken creations. In space, when the rockets misfire, when the O rings and backup O rings fail, when the captain loses his mind, or when the waste-purifying algae develop a disease, as they all must sooner or later, then the story is over. If we could create a truly complete life-support system to sustain us in space, then we would have created the earth.

Comparisons are often made between the dream of space and the dream of Columbus, two totally dissimilar instances. True, the dream of Columbus also failed; he never found a passage to the Orient. Yet it hardly mattered, for instead he found new, living continents and the wealth of the Indies. There is no new world in our part of space, nothing alive, and as for wealth, all we can do in space is spend it on satellites or destroy it with star wars.

For McDougall, who cannot bring himself to cast away the dream, the enemy is technocracy, not technology. At the end of his book he states his case in eloquent and measured terms:

> My instinct tells me that our science and technology, feeble as they are in controlling Nature, are so acute in studying it that they will soon reveal their limits. It is then that man must confess the mortality of his works, without turning on them or himself with contumely. It is then that the orthodox message is a sure guide: God made us, is disappointed in us, but loves us anyway, by which we are redeemed. Technology is our subcreation. We made it, we will be disappointed in it, but we must love it anyway, or it cannot be redeemed. . . . Somehow . . . we must try to compete for power without making power an end in itself, pursue knowledge without mistaking knowledge for truth, cultivate our subcreation yet know its fruits as only part of reality, and remember that for everything we gain there is something lost.

This is an intensely Christian message, and Original Sin looms large in it. We have sinned, but God will redeem us with His grace. Technology sins, and we must redeem it through something akin to grace. What is the grace that we mortals have to offer? A kind of collective and general forbearance and restraint, an exultation in our power combined with a knowledge of its limitations. What will bring this state about? McDougall doesn't exactly say; maybe it will be that same knowledge of our limitations.

The hope is a poignant one, but I will not trust my life or the life of my planet to such an inadequate expectation. First, technology is not as distinct and separate from us as we are from God—the analogy doesn't hold. More important, even if the parallel were good, Original Sin, one of the weakest parts of Christianity, is not the right model for the working out of earth's salvation. If we are inherently evil, what is the point of trying to be good? If technology is inherently evil, if we are bound to be "disappointed" in it, what is the point of our loving it anyway, and what good will that do if we have to live with it? How is that love to remain different from the technology-worship that McDougall describes so well and scorns?

I prefer the Jewish approach to sin as a guide for coping with the dangers posed by technology. Adam and Eve's sin, although it created problems for us all, does not have to be repeated by anyone. Everyone is different. Some choose a good path, others a bad, most a mixture of the two. God's final judgment is on each total performance, and varies from case to case. Love is always available to anyone who manifests a willingness to turn and follow a better direction.

By analogy, all technology is not the same. Some is good, some is bad, some is mixed. We are under no compulsion to love a particular technology, and rejection of one technology does not imply rejection of all others. There is no blanket grace for technology, nor should there be.

A similar approach can be found in Buddhism, which like Judaism is preoccupied with practical, earthly cares, not with Heaven and theological abstractions. This is no doubt why many of the most thoughtful and caring of the modern lovers of technology—Robert Pirsig, E. F. Schumacher, and Arnold Pacey come to mind—

have been inspired by Buddhism. Pacey, in *The Culture of Technology,* writes:

> For anybody who thinks that sensitivity in the use of technology is
> important, even within a rationalist frame of reference, Buddhism
> can present some refreshingly challenging views. . . . There is
> . . . a tendency to think about human creativity more modestly,
> with less stress on virtuosity. . . . There is the suggestion that ideas
> such as charity, sharing and the meeting of basic needs should be
> taken much more seriously as goals in technology.

Pacey's book has examples of technologies that meet these criteria, as well as those, like space and weapons technology that fail. He insists, and I agree, that the fatalistic notion of the steamroller of "technological determinism" is a myth; history demonstrates that we have the freedom to pick and choose among technologies.

And Pacey provides another conclusion, also conspicuously missing from McDougall and from the space story, a conclusion that has probably occurred by now to many readers. In the picking and choosing that must end technological determinism, women have a critical part to play—not as manipulated and exploited images or symbols, but as proven developers and selectors of user-friendly technologies in many parts of the world, especially those places we call "undeveloped." Pacey's most striking example comes from Kerala, one of the poorest states of India, where the health of children and the life expectancy of adults improved greatly over the national average after the widespread introduction of a new technology: teaching people, especially women, to read.

As for space technology, it may well be beyond redemption. In . . . *The Heavens and the Earth,* I came across James Webb's quotation of Raymond Bauer, "whose study of the space program suggested that its effects 'may include changes in man's conceptions of himself and of God.'" This is but an echo of Tsiolkovsky's words, mankind "must surely achieve immortality." Now, after *Challenger* (and after Chernobyl), I hope that fewer will take this sort of *chutzpah* seriously. Rather than rethink our ideas of God, we would do better to reconsider the nature and effects of our own creations. We need not follow the example of the citizens of Babel and bring our language, our "city," our technology, down in ruins upon our heads. There is time to choose to stop building ill-starred towers to challenge the heavens. There is time to pause, and time to find other edifices of less height and greater grandeur upon which to lavish our creative powers.

After Valdez

The oil that leaked from the ruptured hull of the *Exxon Valdez* into Alaska's Prince William Sound on Good Friday, March 24, 1989, had many effects. It provided a windfall income to local residents who rented extra bedrooms at one hundred dollars per night to visiting scientists, administrators, lawyers, and others concerned with the spill and the cleanup. It gave the professional opportunity of a lifetime to anyone willing to put in 110 hours a week, under horrendous conditions, studying oiled biota—everything from whales to coyotes to eagles. And of course it devastated one of the most beautiful ecosystems on earth.

For those of us who call ourselves conservationists, Valdez was, above all, a source of enormous and bitter frustration, a litany of lessons unlearned and opportunities missed. This was not for lack of dedicated, competent investigators on the scene after the accident, nor was it for lack of funds to support their work. The problem was rather one of context. Between the carefully refereed, organized, informed sorts of conservation biology reported in scientific journals and popular magazines and the real world of emergency actions taken in Alaska there is much difference. All too often, biologists who knew what had to be done, at least the first steps, were checked by political and bureaucratic constraints that many of them had not even imagined. Worse yet, actions were taken that might not have been attempted had wiser counsels prevailed. The immediate side

effects and the ultimate consequences of these steps were not considered; in many cases, taking no action at all would have been preferable.

A nightmarish job was made even more difficult because everybody and nobody was in charge. How does one do a proper necropsy
on a dead porpoise—hard enough at midnight in a makeshift dissect-
ing room—if representatives of several federal and state agencies are
competing over who gets which parts of the carcass? With at least
eighteen governmental agencies on the scene, jurisdictional disputes
were frequent, complex, and often bitter. Scientists padlocked their
specimen containers; rival scientists padlocked the padlocks.

But battles over turf were not the only handicaps imposed upon the
hard-working recovery teams. How is it possible to do decent science
in a disaster area, or to understand the results afterward, when the
findings must be kept secret during the critical first year or more
following the accident? The government's environmental agencies
are run by lawyers; they set the priorities. Their first priority in
Alaska was to collect evidence to help win the case against Exxon.
Evidence gathered in a lawsuit must be kept confidential until it is
introduced in court, lest the opposition find out what cards are in your
hand. Even the scientific seminars given by governmental scientists
about the Valdez spill were precensored by the lawyers. Naturally,
the scientists working for Exxon were under a similar injunction of
secrecy. I am unwilling to name the sources of the information in this
chapter for fear that I might damage the careers of those who spoke to
me. There was less secrecy after Chernobyl.

How can the right conservation management decisions be made
when the experienced professionals on the spot are not consulted?
Who decided to fly five oily young harbor seals to Anchorage, where
they were kept in a veterinary facility that also houses dogs and cats?
Who decided to release them back into Alaskan waters after their
recovery? (The epidemic that killed thousands of seals in the North
Sea is thought to have been caused by a virus very similar to canine
distemper.) What was the purpose, anyway, of returning to a nonen-
dangered, wild population five animals that had had such unnatural
contact with human environments?

In the same vein, who decided to rehabilitate oiled sea otters and
to return the approximately two hundred survivors of the treatment to
the sea, at a cost of forty thousand dollars per otter? (Payments to

local fishermen for oil-related losses averaged approximately twenty thousand.) Perhaps it was the right decision, but few, if any, knowledgeable wildlife biologists at the scene were involved in making it. Nor was there consultation with any philosophers familiar with issues in conservation ethics. In these and other critical decisions at Valdez, politics was the guiding influence and public relations replaced informed public debate.

Of all the frustrations arising from the recovery effort, some of the saddest were occasioned by vicious and intractable conflicts among recovery workers. One such incident occurred when two governmental scientists advocated the surgical implantation of radio transmitters into rehabilitated sea otters that were about to be released. Angry local residents employed in the recovery facility, located on a remote peninsula, gave the two women a canoe and paddles and forced them to cross the bay and take refuge in the nearest village in the middle of the night. Subsequently, the transmitters were implanted in the otters without mortality, a fact that may or may not be relevant to the dispute.

The Valdez recovery effort was marked by heroic and self-sacrificing performances by many conservation scientists and managers, including men and women employed by Exxon and related companies such as British Esso. But it was not a triumph for conservation. The simple truth is that *there is no way to be prepared ahead of time to cope with disasters of this magnitude.* Some of our environmental follies are beyond repair even if we spend billions of dollars in the effort. There will be interagency task forces formed, reports written, and eventually, scientific papers published about the Valdez spill. But the next massive oil spill will again find us wondering, years later, whether the cleanup has made the situation better or worse.

The feeling of futility that many of us associate with the name of Valdez can be conveyed by telling, however briefly, about some of the results of the cleanup effort. First the story of a badly oiled male sea otter which was de-oiled and painstakingly nursed back to health. After a long period of convalescence in captivity, the now robust animal was released outside the spill zone in pristine waters that had been identified as proper habitat for sea otters. Nevertheless, its first act after being freed was to swim the ninety miles back to its former home, where it was found dead a few days later, looking much as it had right after the oil spill. Similar stories are told of harbor seals, whose newborn pups are placed by their mothers on the soft vegetation that covers the rocks of their rookeries. This vegetation was given much attention by the cleanup crews, and the oil was carefully removed from its surface. But pups placed on it died anyway, as their weight caused more oil to ooze up from the spongy subsurface layers. Finally, there was the inadvertent, massive disturbance of erstwhile wilderness caused by the activities—the mere presence—of so many people (with their boots, boats, machinery, garbage, toilets) who were there to help out.

The all too evident fact was that after a few days oiled shores that had been "cleaned" looked pretty much the same as those that had not. Nor was the eye deceived in this: the cleanup was often an expensive fiction. Shingle beaches provide a good example—soon after the crews had passed, oil seeping out from under stones and from crevices coated all the rocks again, making them look as if they had not been touched by the cleanup. Indeed, the cleanup was sometimes more destructive than the oil. The hot water used to clean some of the beaches cooked any of the beach organisms such as mussels that had survived the spill. In places where the oil had been deposited on the upper beach like the ring around a bathtub, the attempt to remove it sent cascades of toxic, oily water down to the previously less affected lower beach and subtidal zones.

The *Exxon Valdez* disaster was only the thirty-fourth-largest oil spill in history up to that time, but because it affected us in North America, it caught the public eye. Perhaps we can learn from it. There will be many lessons of Valdez, but the most important one is apparent now: society must scale down its industrial processes, including the long-distance transport of hazardous substances. Only then can conservationists cope with the smaller accidents that inevitably will occur, or, if they can't cope, safely leave the relatively small damaged areas to natural processes of restoration. Such a decrease in the scale of industry is likely to be brought about only if the developed world decreases its material consumption and makes wiser choices of the things it consumes. This is a social process in which expert and layman will have to participate together on equal footing, without secrecy or censorship, as common members of a society at great risk.

Dent de Lion—
The Lion's Tooth

George Orwell wrote that if you want to understand a prejudice, the best place to start inquiring is inside your own head. That's not easy, because before you can get anywhere with your inquiry you have to be able to say to yourself, "I am a prejudiced person. I have an irrational belief or feeling." Only after making this admission do you have a chance of finding out the source of the prejudice.

So I admit it, although reluctantly: I have often picked dandelions out of my lawn. As I push my old-fashioned reel-type lawnmower over the patches of grass that haven't been sacrificed to vegetable garden or fruit trees on my one-eighth-acre estate, I enjoy watching the yellow dandelion heads fly up from the blades, and get grim pleasure from slicing off any erect leaves close to the base of the persistent weed. Dismembering dandelions is fun—one of the few legal and inexpensive pleasures left to us in an overcrowded world beset with bureaucrats, technocrats, and their collaborators. Better yet, women and men, toddlers and ancients, can all partake of this most egalitarian of pastimes and the neighbors will smile and wave their approval.

But even though it's a near-universal ideal in this country, wanting an all-grass lawn is still a prejudice, an irrational desire with all kinds of harmful consequences. What makes it so odd that I should share

99

this prejudice against dandelions is that I know all the harmful consequences of a weed-free lawn; I even make a pest of myself telling the
neighbors about them. Yet I keep on killing the dandelions. Why?

A superficial inquiry into my prejudice would stop at the idea of
social conformity. Weedy lawns are the great American bugaboo.
Flag burning is unpopular, but a few dozen dandelion puffballs will
really arouse the passions of the suburban mob. Let the person who
coddles dandelions in the midst of Oak Grove or Shady Pointe or
Country Terrace Mansions beware: a brick through the plate-glass
living room window is pure routine in such cases, as inevitable as the
delivery of the mail.

I can't deny that I am influenced by the opinions of my neighbors,
yet I doubt that they fully account for my "dandephobia." My part
of town is not truly suburban—the houses are too close together and
too close to the street to permit much show of lawn. Moreover, when
it comes to lawns, some of the people on the block are decidedly
weird. One elderly, conservative neighbor several years ago replaced
all his grass with stones of a prison-yard gray. Most of the smaller
stones have since disappeared into the pockets of passing children,
but the grass has not come back. Down the street, a liberal activist
couple allow their few wispy strands of grass to grow to several feet
in length before borrowing a power mower from the banker five
houses away. And strangest of all are the owners of the house on the
corner, where little grows except after the rain, when my wife and
children rush down the block to find bizarre mushrooms that don't
key out in any of the field guides. You can see why I reject social
pressure as an explanation of my prejudice. To understand it, I must
dig a little deeper.

Embarrassing as it is to admit, I think the problem is that I share
some of this society's passion for control. We have to control everything: our environment, our children's futures, our entertainment,
our looks, even—in the case of artificial hearts—our mortality. Of
course most of this control is ultimately a pretense and a sham; even
when it works, the result is always more modest than the expectation.
Dandelions are the supreme symbol of the failure of human control, a
yellow flag of mockery, and every time we burn that flag, back it
comes, stronger than ever. No plant or animal is as obstinately per-
verse in its flaunting of human wishes.

The dandelion carries its obstinacy to an extreme. It grows where people don't want it, and seemingly nowhere else. One may wander the wilderness of the New Jersey Pine Barrens for hours without seeing a dandelion—if suddenly a few plants appear, chances are there was once a village or a homestead in that spot, long abandoned and vanished save for a few bricks and bottles, a dot on an old map, and the dandelions. Or consider Adrian Wells, of Wilton, Maine, who according to *The New York Times* of June 18, 1981, was the only farmer in the United States growing and canning dandelions for year-round consumption. After gazing out over his second failed dandelion crop of the season, Mr. Wells pulled up a large specimen growing wild in the shade next to his cannery.

" 'Now look at this healthy one; it's beautiful and big enough to feed a horse. . . . What I can't understand,' he added, 'is how these things will grow so well when you don't want them to and then when you put them out in a field with air and sunlight and water and fertilizer and you even weed them they develop every problem known to man.' " The headline on this story, a classic, was: "A Crop Withers for Lack of Adversity." Here is the root of my prejudice—and yours—against dandelions: every dandelion proclaims that the emperor has no clothes.

Biologists know much about the dandelion and can explain many of its strategies for succeeding where it is not wanted. *Taraxacum officinale* Weber, a relative of lettuce, most likely originated somewhere in Europe or Asia Minor. Everything about it adapts it for life in lawns, gardens, and around houses. It is a perennial, enabling some individuals in every population to survive for years and outcomplete more evanescent plants. It has a long tap root—up to three feet—allowing it to reach water and nutrients. Dandelions adapt to an enormous variety of soils, conditions, and climates. The stem is almost nonexistent and the leaves form flat rosettes, allowing the plant to escape the blades of mowers and scythes. Like many weeds, it reproduces asexually, and can thus produce seeds without pollination or a partner. Its life cycle is incredibly short: young plants flower early; the whole process of flower stalk formation and blooming takes three days, and in two more days the seeds are ready to disperse. Between flower bloom and seed formation, the flower stem lowers itself flat against the ground, avoiding grazing animals and mowers. The seeds are numerous, light, easily dispersed by wind, and therefore ubiquitous—all soil contains dandelion seeds, and they can survive for decades if germinating conditions are not right. Most dandelions are "short-day plants," flowering in spring and fall, but a few eccentrics are always ready to bloom in July, ensuring a continuous production of seeds from March or April to December in many parts of the country. And if the dandelion root is cut by an angry gardener, the portion remaining in the ground sends up two to five new dandelions in an attractive ring around the place where the old plant used to be.

Such an extraordinary success story, a tribute to the life force of nature, ought to be an inspiration to us, but instead we take it as a challenge to our omnipotence. Weeds, like love and depression, are a state of mind—even if we want to, it's hard to change our feelings. Most people go after dandelions with heavy artillery, herbicides. And like heavy artillery in a guerrilla war, herbicides cause a lot of random damage but only give the enemy a temporary setback. Dandelion-killing chemicals are not specific: they can and do damage other broad-leafed plants, including ornamental shrubs and valuable trees. If the chemical companies ever found a pesticide that controlled dandelions for more than a few months, they would suppress it at once; the important thing is to keep the customer coming back for more, like a crack addict. On the other hand, the dandelion-killer must work for a while, so that people will continue to buy it. Also, a temporarily weed-free lawn is much more susceptible to insect pests and fungal infections than one with a mix of plant species, and this means more pesticide sales.

Another incidental casualty of herbicide use is human health, especially the health of the hapless teenagers who work for the lawn care companies, but also perhaps for the children who roll in the poison-soaked, green, green grass. Use of 2,4-D, the principal lawn herbicide and a major component of the Agent Orange used in the Vietnam War, appears to be associated with increased risk of lymphoma. One fellow who was spraying my neighbor's lawn for a chemical lawn care company told me that some of the sprays gave him nosebleeds and severe headaches. It hadn't occurred to him to find another job.

Prejudices don't respond well to reasoned arguments: even as I slaughter my dandelions I know they have many uses. *Taraxacum* is

Latinized Greek for "remedy for disorder," and *officinale* indicates its place on official rosters of medicinal plants. The U.S. Pharmacopoeia lists preparations of *Taraxacum*. I don't know if they work. It is eaten as a pot herb, salad green, and, blanched, in lieu of endive. It contains much more vitamins A and C than lettuce, more iron than spinach, and lots of potassium. Many people don't like its bitter taste, but enough do to put it on supermarket shelves every now and then. The Apache and Digger Indians were reported, in 1870, to travel days just to find dandelions to eat. The dried, ground, and roasted roots make a supposedly excellent coffee substitute. The flowers are used to make dandelion wine or are floured and fried in place of the much rarer morel mushroom. Violet dyes can be made from dandelions. Children once used the peeled flower stems to make curls for dolls, in the days before Nintendo.

The most appealing use of dandelions, for those susceptible to appeal, has nothing to do with their intrinsic properties. They add color to a lawn. Lawns probably originated, according to *The Oxford Companion to Gardens,* as "well-maintained fields used . . . as a setting for ornamental trees and shrubs." Daisies, speedwells, and other low-growing herbs were considered to enhance the appearance of lawns. In medieval times, southern European lawns of clover or grass often had flowers mixed in with them to mimic the appearance of flowery meadows. Some medieval lawns were entirely composed of Roman chamomile, a ground-clinging perennial herb which tolerates trampling, has a sweet smell and little daisylike flowers, and is in the same family as—the dandelion.

How much better and easier it would be to imitate nature than to fight it. Our lawns could be a little like the valley of the San Joaquin, in central California, as John Muir first saw it in the spring of 1868: "the floweriest piece of world I ever walked . . . they are . . . side by side, flower to flower, petal to petal, touching but not entwined . . . one smooth garment, mosses next the ground, grasses above, petaled flowers between."

But until we can shake free of the drive to control everything in our world, we will be in no position to imitate anything decent or beautiful in nature. Our world view and the prejudices it spawns condemn us to endless vistas of sterile, expensive, dangerous, and fragile green lawns, free, for this brief moment, of dandelions, violets, and other demons of our own making.

The Technology of
Destruction

Edward Teller was born in Budapest in 1908, the son of successful, nonobservant Jewish parents. Showing an early talent for mathematics, he was trained in that subject and in chemical engineering, but eventually gravitated toward theoretical physics. In 1931 he became a part of the celebrated Göttingen group of physicists and mathematicians, which included such names as Werner Heisenberg, David Hilbert, James Franck, Enrico Fermi, Niels Bohr, George Gamow, and a youthful pair who roomed at the same villa, Paul Dirac and J. Robert Oppenheimer. Driven from Germany by the rise of Hitler and the Nazis, Teller made his way, with Rockefeller Foundation support, to Niels Bohr's institute in Copenhagen, where he shared lodgings and the love of poetry with a German aristocrat and talented atomic physicist, Carl Friedrich von Weizsäcker. In 1935 Teller traveled to the United States and took the first of a series of university appointments.

By the summer of 1939 only a dozen people in the world appreciated the full significance of nuclear chain reactions. It had become plain that there would be no cooperation among them to suppress the development of nuclear weapons. Moreover, Heisenberg and von Weizsäcker, two of the dozen, intended to remain in Germany, where they eventually headed (and perhaps deliberately led astray) Hitler's unsuccessful effort to develop the bomb. On August 2, Teller, with Leo Szilard—another of the twelve—drove out to the Long Island cottage occupied by Albert Einstein. According to Einstein's and Teller's recollections, Szilard handed Einstein the draft of a letter to President Roosevelt urging the need for a secret program of atomic research.

Much later Teller commented that "Einstein only signed his name. I believe that at that time he had no very clear idea of what we were doing in nuclear physics." Einstein agreed with this interpretation. After the war he said to his biographer, Antonina Vallentin: "I really only acted as a mailbox. They brought me a finished letter and I simply signed it." He also is said (by atomic historian Robert Jungk) to have remarked, "If I had known that the Germans would not succeed in constructing the atom bomb, I would never have lifted a finger."

That letter launched both the American atomic bomb project and the long, productive nuclear career of Edward Teller, "Father of the H-bomb." The August meeting with Einstein was his "first atomic assignment." Neither history nor mythology offers us an example of a single human being who has been as successful in creating weapons of mass destruction.

As an active member of the Manhattan Project from 1941, Teller found his way to the laboratory of Enrico Fermi at the University of Chicago and then to the fission and fusion bomb study group led by Oppenheimer at Berkeley.

In 1943 he moved to Los Alamos, New Mexico, still as part of Oppenheimer's group, but he became increasingly dissatisfied with the direction the research was taking. Impressed with the enormously greater power of atomic fusion compared with that of fission, Teller pushed for the early development of the H–bomb. Oppenheimer and the majority disagreed. In 1951 Teller left Los Alamos, first for Chicago and then for the newly created Lawrence Laboratory at Livermore, California, which he had successfully lobbied for in Washington. In the next year the United States exploded Teller's H-bomb, causing a Pacific island a mile in diameter to disappear in a fraction of a second. About that explosion Teller later wrote: "We would be unfaithful to the tradition of Western Civilization if we shied away from exploring what man can accomplish, if we fail to increase man's control over nature." The explosion of the H-bomb brought Teller to a pinnacle of influence and power in the Western World.

In 1954, at the height of the McCarthy era, Teller testified before the Atomic Energy Commission's security board that Oppenheimer, although loyal, was not to be trusted with the "vital interests of this country," because of his failure during the war and immediately afterward to pursue the development of fusion weapons. Unavailable at the time, for security reasons, was a sharply differing account of the controversy. As claimed by Hans Bethe, atomic pioneer, Nobel laureate, and once Teller's supervisor at the Manhattan Project, Oppenheimer's decision to downplay fusion research after the war was based in large measure on some faulty calculations made by Teller himself. Not until the severity of these mistakes was appreciated did Teller and the mathematician Stanislaw Ulam discover, in 1951, a correct way to design an H-bomb.

After his denunciation of Oppenheimer, Teller was shunned by many of his former colleagues in physics and applied mathematics. Yet he maintained both his academic and weapons connections in California—at the University of California, Berkeley, and indirectly at Davis, later at Stanford's Hoover Institution—and always, in one capacity or another, played a leading role at the Lawrence Livermore National Laboratory. In the quarter-century after the fall of Oppenheimer, Teller, although in partial eclipse, nevertheless guided Livermore into a preeminent position as the nation's designer of advanced nuclear weapons systems. And he became a public champion of many nuclear and militantly anticommunist causes—from nuclear arms buildup to atmospheric testing, to civil defense, to nuclear power and other "peaceful" uses of atomic energy, to the Vietnam War. In 1962 Teller wrote: "We cannot be strong unless we are fully prepared to exploit the biggest modern power, nuclear explosives. Nuclear weapons can be used with moderation on all scales of serious conflict. . . . We should be prepared to survive an all-out nuclear attack."

Let me now speak for myself. As a biologist reviewing this brief outline of the scientific career of an atomic physicist, I am struck not so much by what I find as by what is missing. It is as if Teller and I grew up in different universes with different fundamental laws. Or perhaps we have encountered the same universe but with different organs of perception. Although our address has been the same planet, the attributes of our environment that have mattered to each of us are worlds apart. I can illustrate this better with an example— one tiny part of my scientific world that matters and speaks to me.

There are more than three hundred species of often brilliantly colored Euglossine bees in the American tropics; many of them are the sole pollinators of their own associated orchid species. Among certain orchids of the genus *Coryanthes,* male bees gathering scent from one part of the flower become intoxicated by the odor and fall into a bucket-shaped, water-filled lip of the flower positioned imme-diately beneath the scent glands. The only way out for the bee is up a ladderlike structure in a narrow passageway leading into the heart of the orchid flower. When it reaches the male column, a part of the flower shaped to fit the anatomy of this species of bee slips in be-tween the thorax and abdomen and holds it fast. The bee struggles for fifteen to thirty minutes, pushing back a cap covering the sticky pollen sac, which becomes attached to a specific part of the bee's abdomen by a stalk. The bee frees itself. As it flies off, the rush of air dries the stalk, which contracts, reorienting the pollen sac so that it is positioned to be deposited directly on the female part of the next orchid flower that the bee visits.

In the vast panoply of life this interaction is but the merest fraction of a speck; after all, there are at least five million species on earth, each with its unique life history. But that fact in itself is the place to start the comparison of Teller's perception of the world with mine. His scientific world is one of general laws and elementary relation-ships, which when understood can be manipulated, as technology, for desired ends such as the manufacture of nuclear weapons. The scientific thrill comes with the discovery of the basic laws and this thrill is reinforced by the power that rewards the successful technolo-gist. I cannot be sure, but I imagine that to a mind like Teller's *specific* things and *specific* events are just so much clutter until they can be seen as part of a general pattern.

Of course, biologists also look for generality. General laws are important, yet to a real biologist much of the wonder is in the richness, the diversity of life. Thousands and thousands of kinds of plants are pollinated by insects—only *Coryanthes* uses that particular bucket-and-ladder mechanism. Diversity is a primary manifestation of biology, as precious and wonderful to those who study life as the laws of mass and energy are to those who study physics. To destroy such diversity, for any reason, is unthinkable.

Another example will take the comparison a little further.

At coral reefs in all tropical oceans, many divers have observed an extraordinary process taking place. Little fishes, usually wrasses, gobies, or juveniles of other species, set up territories, either singly, in pairs, or in small groups. These "cleaner fish" often have the same color pattern: blue and black longitudinal stripes. Before long, much larger fish assemble at the cleaning station, frequently ordering themselves in patient, mixed-species queues of as many as fifteen to twenty individuals. Assuming special postures and sometimes colors, the large fish at the head of the line are cleaned of parasites by the smaller fish, which scrape off the parasites and eat them. In the course of their work, the cleaners swim in and out of the mouths of the larger fish without harm. After being cleaned, a fish moves peacefully off and its place is taken by the next in line. Many of the clients of these stations are predators accustomed to eating wrasses and gobies when they find them in other situations.

It is natural for observers of other forms of life to find in nature things that seem applicable to human society. This is the so-called anthropocentric approach, and although it has its dangers it can be useful and even inspiring. True, each generation finds the things it wants to find in nature, but eventually a complete picture begins to emerge. Nineteenth-century biologists living in a society in which the working classes were brutalized by a newly emerging, raw industrialism found plenty of examples of subjugation, exploitation, and brutalization in nature. Although this ruthless view of nature still exists, many biologists are now sensitive to the other side of the story. As the reef fish tell us, enemies can cooperate to their mutual benefit even while they remain enemies. Nature abounds with such examples—the more diverse the natural community, the more positive relationships occur within and between species.

Unfortunately, complex living systems with their myriad positive and negative interactions are especially vulnerable to outside interference. The more complex, the more vulnerable, as seemingly small changes resonate destructively throughout the system. When people introduce technology into the natural world, unexpected and unwelcome things can happen.

Pueblo Bonito was a large communal dwelling built by Native Americans in Chaco Canyon, New Mexico, in the year 919 c.e. A success from the start, it was greatly expanded and remodeled in the middle of the eleventh century, becoming a masterpiece of passive solar design. Oriented so that the first ray of sunlight on the morning of the summer solstice struck exactly parallel to the front of the pueblo, the semicircular structure attained near-maximum theoretical efficiency of heat retention in winter and heat loss in summer. Less than one hundred years afterward, Pueblo Bonito was abandoned, for no obvious reason. It is speculated that the deforestation occurring when timber was taken from the canyon for the expansion of the pueblo caused soil erosion, an increase in runoff of rainwater, a slow drawdown of the water table, and ultimate failure of the pueblo's agriculture and water supply.

The Pueblo Bonito was an outstanding success: it offered its residents the chance for a stable and probably comfortable existence. But success spoiled it. New Pueblo Bonito was technologically more sophisticated; however, it was out of scale with its natural environment. The late René Dubos, microbiologist, philosopher, and environmentalist, observed that many natural systems are totally unsuited for receiving sudden inputs of large amounts of energy. This is especially true when the energy is in a form that the community has not previously experienced. All successful technologies consciously or unconsciously recognize this principle. Technological cleverness has to be accompanied by restraint or it destroys life, including the life that created it.

What brought Edward Teller back into the forefront of national weapons policy, even as a sick man in the twilight of his career, was Star Wars. Indisputably this was a Teller creation, nurtured for decades in the fastness of the Lawrence Laboratory and finally brought to fruition by the bright young men of the Department of Applied Science. Fellowship support for these bright young men came from the little-known Hertz Foundation, whose official address was a post office box in Livermore, and one of whose directors was—Edward Teller.

They were the bright young men (scarcely any women) of "O-Group," led by the bearlike, combative Lowell Wood, Teller's protégé and intellectual son, spewing futuristic weapons ideas like sparks from a welder's torch. And there was Peter Hagelstein, inventor of the nuclear X-ray laser—medical dream turned into weapons dream. Also Rod Hyde, MIT graduate at age nineteen, who came to Livermore to design a starship engine that might help him leave this limited planet. And others with other devices to invent microwave weapons, particle beams, electromagnetic pulse weapons and those devices perhaps most important of all, that we don't know about yet, all sharing the commonality of unimaginable power.

To Teller and his disciples, Star Wars offered the hope of a world free from the nuclear shadow. Opponents have countered that if so, it would be the first time in human history that any technologically advanced defensive weapon has proved impregnable or has conferred a lasting advantage, and the first time that such a technological advance has not led quickly to a permanent increase in the destructiveness of war. Nor, they have asserted, is there any scientific reason to expect that Star Wars will prove an exception.

It was the X-ray laser, Teller's apparent favorite and O-Group's major announced success, that convinced President Reagan in March 1983 to accede to Teller's promptings and opt for Star Wars. But paradoxically, it was the same nuclear-bomb-pumped X-ray laser that alerted both the public and Washington to the fact that Teller's antinuclear defense was itself nuclear powered. Whether the delivery system is "pop-up" or orbital, X-ray lasers are likely to mean nuclear bombs going off above our atmosphere. Furthermore, X-ray lasers have offensive capabilities. And finally, nobody but Reagan has ever believed that devices such as X-ray lasers could stop all incoming missiles, let alone nuclear bombs, which are perhaps already of convenient carry-on size and capable of being smuggled into the country in suitcases. This awareness constitutes the biggest cloud (apart from the titanic cost) in the Star Wars firmament and the greatest challenge to Teller's vision.

There is no way of telling exactly what a nuclear war would do to life on earth; all the supercomputers in existence could not begin to keep track of the variables that might matter. However, the impact of so much concentrated energy on so many parts of the biosphere could not be other than catastrophic. Vietnam is still devastated by the ecological consequences of its limited, nonnuclear war, and ecologists estimate that extremely serious effects will persist on a time scale measured in hundreds of years, perhaps in millennia.

Here is one scenario of the effects of nuclear war on the living planet, prepared by physicists and biologists working together:

> Extreme cold, independent of season and widespread over the earth, would severely damage plants, particularly in mid-latitudes in the Northern Hemisphere and in the tropics. Particulates obscuring sunlight would severely curtail photosynthesis, essentially eliminating plant productivity. Extreme cold, unavailability of fresh water and near-darkness would severely stress most animals, with widespread mortality. Storm events of unprecedented intensity would devastate ecosystems, especially at margins of continents. . . . Light reductions would essentially terminate phytoplankton productivity, eliminating the support base for many marine and freshwater animal species. . . . Extreme temperatures and low light levels could preclude virtually any net productivity in crops anywhere on earth. Supplies of food in targeted areas would be destroyed, contaminated, remote or quickly depleted. . . . Survivors of immediate effects . . . would include perhaps fifty to seventy percent of the earth's population. . . . Societal support systems for food, energy transportation, medical care, communications and so on would cease to function.
>
> Paul Ehrlich, et al.
> *"Long-Term Biological Consequences of*
> *Nuclear War,"* Science, *December 1983*

Whether this prospect is exaggerated or understated makes no difference—that nuclear war would cause some kind of major disruption of nonhuman life and human society is manifest to nearly every informed person who understands ecological systems and has no stake in the nuclear trade.

And what are we to make of the exceptional career of Edward Teller? Life scientists confront the immense and glorious complexity of life each day. Physicists, by contrast, deal with systems of stark simplicity. Moreover, in living systems irreversibility and uniqueness are ever-present conditions of existence. Entire classes of objects and even processes disappear, never to return again. This matters.

In physics such things are rare. The physicist observing a bubble chamber sees evidence of two elementary particles that collide. The particles disappear and are replaced by another kind of particle with a different mass and maybe energetic waves of certain lengths and frequencies. The old particles are gone, but it doesn't matter. If we wait a few minutes, weeks or months, we will see the identical process happening again. There is no cause for concern about the loss of the original two particles. Perhaps working with these sorts of physical systems is what keeps the Tellers of this world from coming to grips with the peculiar fragilities of life and the human meaning of its destruction.

As a Jew, Teller was heir to a legacy of unnumbered persecutions; indeed in his own lifetime he personally experienced the nightmare of the Holocaust and the iron claw of the Soviet beast. For millennia his ancestors have suffered persecution and enjoyed fantasies of escape—even of revenge. But it is not recorded that any of them in their wildest dreams before now considered that it might be worth risking the world in the struggle against oppression.

The issue is not whether we are to abandon our miraculous cleverness, but how we are to allow ourselves to use it. As the great social critic Wendell Berry wrote, "The use of the world is finally a personal matter, and the world can be preserved in health only by the forbearance and care of a multitude of persons."

Ultimately, the biological and the theological perspective fuse into one, and we come to understand that certain acts of destructive creativity cannot be justified or excused for any human reasons. This was the message of the Catholic bishops to those who worked on weapons of mass destruction, whether "offensive" or "defensive." It is a message of the Koran: "Have they never journeyed through the land and seen the fate of those who came before them? . . . And to them, too, their apostles came with undoubted signs, which were rejected to their own destruction. Allah did not wrong them but they wronged themselves." The message is central to the other great faiths as well. And the reason why it is true, as in the case of all important truths, is very simple.

"Man remains a partner of God in the ongoing creative process," suggests Norman Lamm, a Talmudic scholar.

> However, here we must distinguish between two Hebrew synonyms for creation: *beriah* and *yetzirah*. The former refers to *creatio ex nihilo* and hence can only be used of God. The latter describes creation out of some preexistent substance, and hence may be used both of God (after the initial act of genesis) and man. God has no partners in the one-time act of *beriah* with which He called the universe into being, and the world is, in an ultimate sense, exclusively His. He does invite man to join Him as a co-creator, in the ongoing process of *yetzirah*. Hence, man receives from God the commission to 'subdue' nature by means of his *yetzirah*-functions; but because he is incapable of *beriah*, man remains responsible to the Creator for how he has disposed of the world.

This statement is an echo of a much earlier rabbinic teaching recorded in *Ecclesiastes Rabbah*, a text redacted in approximately the eighth century C.E.:

> In the hour when the Holy One Blessed Be He created the first
> man,
> He took him and let him pass before all of the trees of the garden
> of Eden,
> And said to him:
> See My works, how fine and excellent they are!
> Now all that I am going to create for you I have already created.
> Think about this and do not corrupt and desolate My world;
> For if you corrupt it, there will be no one to set it right after you.

Any technology—whether of creation or destruction—has as its first responsibility the preservation of the human and nonhuman richness of the world. It is an old idea, at least as old as the book of Deuteronomy, that even in wartime we have an obligation to leave the trees standing in the cities of our enemies. If to some this obligation seems unnatural, it is a sign of how far we have strayed from the path, and the peril that we are in.

Hard Times for
Diversity

Biological riches are the most dependable kind of wealth. Money, if it ever appears, is easily lost. Possessions are equally vulnerable. A good name takes effort to maintain—it can be erased by a momentary carelessness or, in this era of computer files and insatiable news media, by the carelessness of others. Art and architecture are destroyed by war and vandalized by time. Health vanishes; happiness can disappear like silver maple leaves at the approach of winter. Nobody is proof against such changes. But the diversity of living species and communities, even when locally diminished, remains, a cup that fills itself at every opportunity, although not always with the same wine. One would think that the value of natural and cultural diversity would be obvious to each new generation. Not ours.

The appreciation of diversity has fallen on hard times in Western and Western-influenced societies. The spirit that moved Botticelli to incorporate at least thirty different species of plants in his canvas entitled *Spring,* painted in 1478, the spirit that prompted Shakespeare to mention enough animal and plant species in his plays to provide source material for entire books, and the spirit that gave Thomas Jefferson delight in compiling his great collection of comparative vocabularies of fifty different American Indian languages, if it has not vanished, has been effectively sublimated and suppressed in our day. We have in its place the celebration of the sort of pseu-dodiversity that one finds on the menu of a Chinese-Polynesian-American restaurant, the ersatz variety that is the hallmark of our homogenized world culture.

What has caused the decline of the love of diversity and is causing the decline of diversity itself is, not surprisingly, the ascendancy of its opposite: uniformity. We have abandoned our fascination with the specific, with species, in favor of a preoccupation with the general and the generalizable, with scientific laws. This is the Age of Generality, and every month that passes sees it more firmly entrenched as the official way of seeing and dealing with the world.

Generality and Power

Why is generality in the ascendancy? There is one reason, I believe, and it is all but irresistible. Generality confers power. Much of our control (influence is a more accurate word) over the external world is related in one way or another to our discovery of general laws and principles of physics, chemistry, and biology, an explosion of knowledge that is essentially modern. Our ability to manipulate the world, which goes beyond the dreams of the alchemists, is new and addictive. Never mind that the control we claim is almost always flawed, incomplete, and even self-destructive: we humans are poor at giving up power once we get a taste of it. Of course knowledge of the specific is also useful; but it doesn't very often help us control the world. Penicillin was discovered because of specific observations, but antibiotics as a class belong to the category of applied general theory.

A good example of the power of generality, certainly the one closest to my own experience, is the drastic change that has occurred in biology since 1965. In the modern scientific-technical process, the specific is usually subordinate to the general and is considered the work of lesser scientific intellects and technicians. We expect this mindset of physicist and mathematicians—not of biologists. Yet the "cutting edge" in biology is now popularly thought to be genetic engineering, a science and a technology that take advantage of the uniformity of the genetic code to toss all organisms into the same grab bag of genes. Most of the genes in this bag are considered trash; a few are withdrawn to be inserted as supposedly useful parts into a small number of recipient organisms such as milk cows or wheat. Here is generality taken to the last degree. Although genetic engineering often does not work as well as its glittering promises, the proof of its power is that the genetic engineers now command higher salaries and more political clout than any previous generation of Western biologists.

It cannot be argued that these biologists have simply shifted the level of specificity from the living organism to the gene. It is the organism that we perceive, that lives in our environments and our history. How can we respect a new view of life that treats plants and animals as mere holders or packages of genes that can be blended at will with genes from other packages? The sanctity of diversity, of separateness, is an ancient belief embraced by people who lived as part of nature and experienced its diversity daily. This belief is no doubt a reason for the biblical prohibitions against sowing several kinds of seeds together in the same field, against weaving wool and linen together in the same cloth, and against interbreeding different species of animals. In the words of the Hellenistic Jewish commentator Flavius Josephus, "Nature does not rejoice in the union of things that are not in their nature alike." How alien this concept seems in today's homogenized, nature-distanced, power-loving society. The sanctity of diversity loses all meaning when designer organisms can be patented and sold to the highest bidder like a new type of cigarette lighter or a new kind of toilet seat.

Generality, as practiced by the genetic engineers, has almost completely taken over the one-time shrine of diversity, the science of biology itself. No longer is diversity celebrated for its own sake. No longer do we see, for example, graduate students who have staked out as their primary life's work a taxonomic group, such as mayflies or top-minnows or the birch family; at least we don't see very many of them. Now their work is organized around general questions and themes, such as "foraging strategies" and "habitat fragmentation." Useful as many of these concepts are, they are no substitute for knowledge of the specific. Yet without generality, at least as a cover, few biologists can hope to find employment and support. As I indicated, any academic biologist in any large university can testify to the sheer political power of that most general of all modern biological discoveries, the universal genetic code. So powerful, indeed, is this particular generality that it has all but obliterated the once-prized distinction between pure science and technology, between the academic and the industrial worlds. In the major universities, professorships and even whole departments are now commonly supported by chemical and pharmaceutical companies. But unlike the days prior to the 1980s, when such support had no formal strings attached, it is now contractually contingent on the corporation's first claim to any discoveries and patents that come from the professor's or department's research. No wonder there is a scarcity of top-ranked graduate students who are committed, for instance, to the study of the bats per se, and precious few jobs for those who are, except maybe in the field of bat control?

In a world bent on manipulation and control, the study of specificity is too slow and tedious, and the maintenance of diversity is too expensive, to be tolerated by the people who make decisions. Both the study and the appreciation of the particular, of diversity, take time, and time is money. The trouble with differences and particularities, from the economic view that prevails today, is that there is no general rule for coping with them, except, perhaps, to ignore them.

A fundamental problem and paradox of conservation arises from our preoccupation with the general and from the power that an understanding of general laws brings: we are committed to the kind of exploitative approach to nature that places diversity in jeopardy. This is truly a Catch-22 situation, or, to use Gregory Bateson's handy phrase, a double bind. Our new-found love affair with generality and general laws, and the resulting power to change the earth have enabled us to destroy biological diversity at an astonishing rate, but they have simultaneously caused us to lose interest in the specific and respect for the people who study it. And that prevents effective conservation.

The Value of Species

As my philosopher friends would say, nature in its separate elements has lost its intrinsic value to people, and now has only instrumental value; we value it for its usefulness. In the long run, this insistence on claiming a general, instrumental need and value for every last species and variety is only going to get us into trouble. But it does not occur to us that nothing forces us to confront the process of destruction by using its own uncouth and self-destructive premises and terminology. It does not occur to us that by assigning only instrumental value to diversity we merely legitimize the process that is wiping it out, the process that says, "The first thing that matters in any important decision is the magnitude of the dollar costs and tangible benefits." People are afraid that if they do not express their fears and concerns in this language they will be laughed at, they will not be listened to. This may be true (although having philosophies that differ from the established ones is not necessarily inconsistent with the power to effect change). But true or not, it is certain that if we persist in this crusade to determine value where value ought to be evident, we will be left with nothing but our greed when the dust finally settles. I should make it clear that I am referring not just to the effort to put an actual price on biological diversity but also to the attempt to rephrase the price in terms of a nebulous survival value.

A concrete example that calls into question this evaluating process comes immediately to mind. I came across it in a paper published in the *Journal of Political Economy* in 1973 by Colin Clark, an applied mathematician at the University of British Columbia. That paper, which everyone who seeks to put a dollar value on biological diversity ought to read, is about the economics of killing blue whales. The question was whether it was economically advisable to halt the Japanese whaling of this species to give blue whales time to recover to the point where they could become a sustained economic resource. Clark demonstrated that in fact it was economically preferable to kill every blue whale left in the oceans as fast as possible and reinvest the profits in growth industries rather than to wait for the species to recover to the point where it could sustain an annual catch. He was not recommending this course—just pointing out a danger of relying heavily on economic justifications for conservation in that case.

In the long run, basing our conservation strategy on the economic value of diversity will only make things worse, because it keeps us from coping with the root cause of the loss of diversity. It makes us accept as givens the technological/socioeconomic premises that make biological impoverishment of the world inevitable. If I were one of the many exploiters and destroyers of biological diversity, I would like nothing better than for my opponents, the conservationists, to be bogged down over the issue of valuing. Economic criteria of value are shifting, fluid, and utterly opportunistic in their practical application. This is the opposite of the value system needed to conserve biological diversity over the course of decades and centuries.

Value is an intrinsic part of diversity; it does not depend on the properties of the species in question, the uses to which particular species may or may not be put, or their alleged role in the balance of global ecosystems. For biological diversity, value *is.* Perhaps nothing more, and certainly nothing less. No cottage industry of expert evaluators is needed to assess this kind of value.

Having said this, I could stop, but I won't, because I would like to say it in a different way.

Assigning value to biological diversity involves two practical problems. The first is a problem for economists: it is not possible to figure out the true economic value of any piece of biological diversity, let alone the value of diversity in the aggregate. We do not know enough about any gene, species, or ecosystem to be able to calculate its ecological and economic worth in the larger scheme of things. Even in relatively closed systems (or in systems that they pretend are closed), economists are poor at describing what is happening and terrible at making even short-term predictions based on available data. How then should ecologists and economists, dealing with huge, open systems, decide on the net present or future worth of any part of diversity? There is no way to assign numbers to many of the admittedly most important sources of value in the calculation. For example, we can figure out, more or less, the value of lost revenue in terms of lost fisherman-days when trout streams are destroyed by acid mine drainage, but what sort of value do we assign to the loss to the community when a whole generation of its children can never enjoy the streams near home, can never experience home as a place where they would like to stay, even after it becomes possible to leave?

Moreover, how do we deal with values of organisms whose very existence escapes our notice? Before we fully appreciated the vital role that mycorrhizal symbiosis plays in the lives of many plants, what kind of value would we have assigned to the tiny, threadlike fungi in the soil that bring essential minerals to plant roots and are nourished by the roots in turn?

And there is one more corollary of our ignorance, a corollary that takes us to the tropical forest. It is a given of the modern conservation movement that we can earn money to save rain forests by selling renewable products harvested from them by native peoples. Yet when we calculate the value of the chemicals, fruits, and nuts that we extract from the forest to obtain money for conservation, how do we factor in the very real costs of the changes in the forest brought about by our efforts? We know that extraction of forest products can drastically reduce the species being harvested and all the other species that depend on them, whether or not the extraction is done by natives using low-energy technologies. It is hard to protect endangered species and habitats from legal and illegal demand once the world market is stirred up—even if it is stirred up in the name of conservation. Nor can we predict in advance what will happen once the commercial genie is out of the bottle. It is easier to develop value than it is to calculate the effects of our valuing. Whether it is nuts from the rain forest, ecotourism, or sea turtle farming, I question whether a lasting human relationship with the environment can be based on the premise that conservation should pay for itself so that we don't ever have to limit our desires.

But while building a case against assigning or developing value in wild species and ecosystems in order to save them, I want to leave a very small loophole. Diversity being what it is—diverse—there are no exception-free laws for saving it. Every endangered species or community requires its own unique plan of rescue (which is why conservation is so expensive), and a small number of these plans may legitimately pivot around the market value of species. Separating the few authentic examples from the many spurious ones is a job for ecologists and economists working together.

There is a second practical problem with assigning value to biological diversity. In a chapter called "The Conservation Dilemma" in my book *The Arrogance of Humanism*, I discussed the problem of what I call nonresources. The sad fact that few conservationists care to face is that many species, perhaps most, probably do not have any conventional value at all, even hidden conventional value. True, we cannot be sure which particular species fall into this category, but it is hard to deny that a great many of them do. And unfortunately, the species whose members are the fewest in number, the rarest, the most narrowly distributed—in short, the ones most likely to become extinct—are obviously the ones least likely to be missed by the biosphere. Many of these species were never common or ecologically influential; by no stretch of the imagination can we make them out to be vital cogs in the ecological machine. If the California condor disappears forever from the California hills, it will be a tragedy. But don't expect the chaparral to die, the redwoods to wither, the San Andreas Fault to open up, or even the California tourist industry to suffer—they won't.

So it is with plants. We do not know how many species are needed to keep the planet green and healthy, but it seems very unlikely to be anywhere near the more than quarter of a million we have now. And if we turn to the invertebrates, the source of nearly all biological diversity, what biologist is willing to find a value—conventional or ecological—for all six hundred thousand plus species of beetles?

I don't deny the real and frightening ecological dangers the world is facing; rather, I am pointing out that the danger of declining diversity is best seen as a separate danger, a danger in its own right. Nor am I trying to undermine conservation; I would like to see it find a sound footing outside the slick terrain of the economists and their philosophical allies.

If conservation is to succeed, people must come to understand the inherent wrongness of the destruction of biological diversity. This notion of wrongness is a powerful argument with great breadth of appeal to all manner of personal philosophies. Those who do not believe in God, for example, can still accept the fact that it is wrong to destroy biological diversity. The very existence of diversity is its own warrant for survival. As in law, long-established existence confers a strong right to a continued existence. If more human-centered values are still deemed necessary, there are plenty available—for example, the value of the wonder, excitement, and challenge of so many species arising from a few dozen elements of the periodic table.

And to countenance the destruction of diversity is equally wrong for those who believe in God, because it was God who, by whatever mechanism, caused this diversity to appear here in the first place.

Diversity is God's property, and we who bear the relationship to it of strangers and sojourners have no right to destroy it. There is a much-told story about the great biologist, J. B. S. Haldane, who was well known to be an atheist. Haldane was asked what his years of studying biology had taught him about the Creator. His rather snide reply was that God seemed to have "an inordinate fondness for beetles." Well why not? As God answered Job from the whirlwind in the section of the Bible that is perhaps the most relevant to biological diversity, "Where were you when I laid the foundations of the earth?" Assigning value to that which we do not own and whose purpose we cannot understand except in the most superficial ways is the ultimate in presumptuous folly.

Erwin Chargaff, ironically one of the forebears of molecular biology, including genetic engineering, remarked not too many years ago, "I cannot help thinking of the deplorable fact that when the child has found out how its mechanical toy operates, there is no mechanical toy left." He was referring to the direction taken by modern scientific research, but the problem is a general one, and we can apply his comment to the conservation of species as well. I cannot help thinking that when we finish assigning values to biological diversity, we will find that we do not have very much biological diversity left.

To Preserve Diversity

Conservation is inextricably linked to human values—linked in its methodology and linked in its chances of success. Until science and society regain a fascination with diversity, with differences, with uniqueness, and with exceptions, all in their own right, there will continue to be a shortage of taxonomists, there will continue to be new and faster methods of cutting down tropical forests, there will continue to be an accelerating loss of species and communities, despite all the science, land, and money that conservation can muster.

True, we cannot unlearn exploitative technologies. Nevertheless, the prevailing human value systems can change in response to need and unknown factors, sometimes surprisingly quickly, and perhaps not by any deliberate acts of individuals. The love of diversity is now under a cloud, but it has not gone away—we would have no zoos, botanical gardens, or natural history museums if it had. The world, I believe, is in the process of discovering that the disastrous effects of exploitative generality can be curbed and moderated by a judicious application of diversity of all sorts (maybe including a growing diversity of nations). Whether we refer to the failures of global economic and political systems, of endless forest monocultures, or of vast irrigation and hydroelectric projects, the cure resides in the various correlates of diversity: local or here rather than widespread or elsewhere, smaller rather than larger, many ways rather than one way, slow rather than fast, personal rather than impersonal, particular rather than general. And this is another paradox, but a pleasant one: diversity does have a general instrumental use after all. It is not that five or ten million species are all ecologically necessary to keep the biosphere intact; but if we relearn how to value this diversity for its own sake, we will discover that we are no longer destroying the world.

In *The Natural History of Selborne,* Gilbert White began his fifth letter to Thomas Pennant with the phrase "Among the singularities of this place. . . ." It was the genius of White that immortalized such specificities as the way golden-crowned wrens hang from branches, the number of spines in the dorsal fin of a loach, or even the diameter, in yards, of the pond near his town. However, it required more than White's genius to make the book so popular for two hundred years. It required the human ability of his readers to value and prize the singularities of a place. That trait still exists but lies dormant in most people, and can be awakened. One of our prime tasks is to find ways of arousing it again.

In my optimistic moments, I look forward to a world where the genius of a Gilbert White or a Linnaeus can thrive alongside the genius of a Watson and a Crick. This isn't nostalgia, such a world has not existed before. It will demand of us one of the most creative advances of recent human history. Those who will bring about the necessary change are the ones who still love diversity for its own sake.

III

True Heading

Once meek, and in a perilous path,
The just man kept his course along
The vale of death.
Roses are planted where thorns grow,
And on the barren heath
Sing the honey bees.

WILLIAM BLAKE
The Marriage of Heaven and Hell

Rights

As the last decade of our century and millennium rolls on, I find myself describing the 1900s in terms of book titles. *The Age of Mega-War*. *The Century of Technology*. *The Era of Environmental Destruction*. *The Epoch of Information*. *The Period of Centralized Bureaucracy*. If there are still universities a hundred years from now, and if they still have graduate students in history, then here are five good, late twenty-first century dissertations in the making—unless somebody grabs them first. I bequeath them freely and without obligation. Lest that be considered a cheap gift of shopworn and stale ideas, I will, sixth and lastly, throw in one that is slightly less obvious: the twentieth century will be known as the Age of the Discovery of Rights. But *The Age of Rights* makes a snappier title.

There is no way to say exactly when the Age of Rights began. Certainly it was not at the stroke of midnight that brought the nineteenth century to an end. The beginnings were earlier and later than that. It began in 1879 in a darkened theater in Copenhagen, when Nora Helmer announced to her dazed husband, Torvald, "I believe that before all else I am a reasonable human being, just as you are— or, at all events, that I must try and become one." It began in 1916, when the U.S. Congress first tried (two years later it was rebuffed by the Supreme Court) to limit child labor through its ability to regulate interstate commerce. It began in 1917, when John Muir lamented that there was "no recognition of rights" for walruses, "killed oftentimes for their tusks alone, like buffaloes for their tongues, ostriches for their feathers, or for mere sport and exercise.'' It began anew in 1949, when Aldo Leopold posthumously announced to the world that the land and the animals and plants that live on it have a right to a ''continued existence in a natural state.''

As Leopold pointed out, there has been an evolutionary extension of rights since the time when godlike Odysseus returned home after many years of fighting and wandering, and ''hanged all on one rope a dozen slave-girls of his household whom he suspected of misbehavior during his absence.'' But only in our century has the evolution of rights become an explosion. Sensing what was coming with uncanny foresight, Samuel Butler devoted a chapter of his satire *Erewhon,* published in 1872, to ''the rights of vegetables,'' an idea that he considered preposterous. Yet one hundred years later, Christopher Stone, a legal scholar, argued in *Should Trees Have Standing?* that trees, rivers, and other components of the natural world, suitably represented, ought to be able to sue in court those individuals and organizations that damage them. Only now, after a century of the discovery of new rights, are we beginning to wonder whether, like limits to growth, there are limits to rights, and whether the damage caused in some cases by the extension of rights might be greater than the injuries the rights were intended to address.

Take the case of Wrightson Island, an unmapped, six-square-mile chunk of land in the Indian Ocean, about seven hundred miles from the nearest mainland. Prior to its first recorded visit by humans, the only species of special note living on the island (there were also bacteria, lower plants, a few fungi, insects, and some sparsely occurring grasses) were a tree, *Arcania australis,* in the guava family, and a bird that nested in it, the Wrightson giant sea sparrow. Young sea sparrows, born in October, grew rapidly on a regurgitated pap of algae and beetles mixed with the fleshy pulp of the *Arcania* fruits, fed by both parents. Neither the bird nor the tree existed anywhere else in the world.

Then, on January 21, 1611, the great Portuguese explorer Juan da Goma, known as Juan the Disoriented, lost because of a malfunctioning sextant and terribly seasick, made an unexpected but welcome landfall on Wrightson Island. Da Goma's log showed that he stayed a week, tinkering with his sextant and resting on terra firma under an *Arcania* tree, while his crew vainly tried to recapture the ship's pregnant goat, which had escaped from a temporary corral on the island. "The gyrfinches are everywhere," wrote da Goma, "very tame and amiable. The flesh is horrible in taste, rancid and bitter. My men soon tire of stomping them for sport."

Three hundred and seventy-four years later, an International Wildlife Fund expedition to Wrightson Island reported a sad state of affairs. Da Goma's pregnant goat had become approximately 380 goats. All grasses and low vegetation were gone, grazed by the voracious goats. No tree seedlings or saplings were found; the age of the youngest of the surviving *Arcania* trees, determined by boring a core and counting the rings, was 327 years. There were only 52 sea sparrows left, including juveniles. The IWF's conservation strategy recommended immediate extermination of the goats by sharpshooting, the only practical way to remove them, and a return voyage was planned. The proposal to shoot the goats soon ran into fierce opposition, however, from a militant group of British animal rights activists known as the Mammal Liberation Front. The IWF office in Liverpool was bombed and the return expedition to the island was cancelled.

In this kind of situation, what guidance do we get from an analysis of rights? The trees have rights; the birds have rights; the goats have rights. Do the goats have lesser rights because they are only a tiny fraction of existing goats in the world, or because they "don't belong" on Wrightson? Are the rights of the trees and birds weakened because there are now 150 healthy specimens of *Arcania* growing well in botanic gardens around the world and large breeding colonies of giant sea sparrows in several zoos? Rights are not supposed to be bent or shaped to fit situations. In this kind of dilemma there are no satisfactory answers based on rights. When rights clash with rights, the ethical path is soon lost.

As rights proliferate, conflicts multiply. Some of them are momentous and dreadful, for example, the struggles in our society over the rights of criminals versus the rights of victims, or the rights of doctors and nurses versus the rights of AIDS patients. This is not to say that rights shouldn't be, that there are too many rights, but that rights are not adequate to all judgments. We need a way of discriminating among rights, of finding out which right takes precedence when rights clash. The answer cannot be determined by examining the rights themselves. It must come from outside, from a larger system that transcends the viewpoint of the individual.

Conflicts of rights against rights can only be resolved by invoking larger entities such as nature, society, God, or some combination of them. To do so moves us away from the modern, individual-centered realm of rights into the distinctly old-fashioned, sometimes authoritarian realm of obligations. Meeting these obligations may damage individuals, but if the obligations are not met the resulting chaos violates far more rights. If the goats on Wrightson Island are not removed, the *Arcania* trees will gradually die out, and the birds with them. Then most or all of the goats will starve as the last of the vegetation disappears and the rootless soil blows away in the wind. Wouldn't it be better to accept an obligation—in this case to maintain the integrity of nature—and to rectify, if possible, the damage originally done by human beings by eliminating the goats?

For all of us, and especially for environmentalists, life is or should be a tightrope walk between the realm of rights and the realm of obligations. Tilt too far to either side and the results will be unpleasant. Too many rights in conflict and the fabric of the larger system shreds and disintegrates. Too many obligations, mistakenly assumed and self-righteously imposed, and great harm can be done in the name of justice.

One example of the mistaken assertion of rights may be the attempt to protect kangaroos in Australia from all exploitation, which then pits the sheep rancher against the kangaroo, a grass-eating "pest." If there were controlled killing of kangaroos for their low-cholesterol meat and fine leather, they might well become more valuable to ranchers than sheep, fewer sheep would be raised, and the outback would recover from the very destructive effects of these hard-hooved, close-cropping, alien animals. Of course it is possible that by assigning a monetary value to kangaroos we will run into some of the problems described in the previous chapter. This is one of those cases where ecologists and economists will have to collaborate.

On the other hand, an example of a mistaken obligation was the decision of the Canadian government to destroy the entire buffalo herd of Wood Bison National Park in order to prevent the spread of bovine tuberculosis to nearby cattle, and to replace it with a genetically inferior herd. There is evidence that wolves can do a satisfactory job of culling the sick animals without our interference; besides, the real culprits appear to have been the game ranches, which can act as conduits for the spread of disease and undesirable genes into once-healthy populations of bison, elk, and deer.

Countless places, not just islands, have problems similar to those found on Wrightson. Many of our national seashores, parks, and wildlife refuges are plagued by introduced horses, donkeys, pigs, and other "exotic" animals. "Plagued" is a judgment, but these feral animals are doing great damage to native vegetation and to native species of animals (for example, introduced pigs on Cumberland Island National Seashore eat sea turtle eggs and hatchlings), and it is a judgment that I don't mind making.

Susan Bratton, a senior scientist with the National Park Service, made the same judgment: in the 1988 volume of *Conservation Biology,* she proposed a list of criteria that can help reserve managers determine when to remove exotic animals. In this list she stated unequivocally: "The burden of proof should rest on the advocates for feral species not on the advocates for native species." In other words, we must shoot Wilbur the Pig if he won't stop digging up the nests of Sally Sea Turtle. I doubt that Dr. Bratton has decided that sea turtles and live oaks have greater inherent rights than pigs. What she seems to be saying is that she and the National Park Service have an overriding obligation to protect the natural systems they are in charge of and to keep them from being unduly altered by human actions, including the human tendency to drop alien pigs, horses, and donkeys around the landscape.

The evolution of rights has been good for conservation, but the assertion of rights is not sufficient to protect the earth. If the twentieth century has been the Age of Rights, perhaps the twenty-first will be the Age of Rights and Obligations—at least I hope so. Neither way of dealing with the world is adequate by itself.

Loyalty

A corporation is an artificial being, invisible, intangible, and existing only in contemplation of law. Being the mere creature of law, it possesses only those properties which the charter of its creation confers upon it, either expressly or incidental to its very existence. . . . Among the most important are immortality, and, if the expression may be allowed, individuality; properties, by which a perpetual succession of many persons are considered as the same, and may act as a single individual. . . . It is chiefly for the purpose of clothing bodies of men in succession with these qualities and capacities that corporations were invented and are in use. By these means, a perpetual succession of individuals are capable of acting for the promotion of the particular object, like one immortal being. . . .

CHIEF JUSTICE JOHN MARSHALL, for the majority;
Trustees of Dartmouth College *v.* Woodward, 1819

With these elegant words, the Supreme Court of the United States established early in our history the legal principle that groups of people charged with specific objectives such as teaching, managing national forests, or simply making money can be vested with a kind of collective existence by articles of incorporation or by acts of legislature, and that this existence can be maintained indefinitely by the successors of the original members of each group. Henry Ford, Thomas J. Watson, and Gifford Pinchot are dead, but the Ford Motor Company, IBM, and the U.S. Forest Service live on. A few strokes of the chief justice's quill pen made legal fictions come alive—gave them not just individuality but immortality in the eyes of the law and the eyes of the generations of Americans to follow.

It is now second nature for us to treat corporations and similar organizations as if they were people: we like them, hate them, honor them, respect them, despise them, and sue them just as if they were living persons with character, personality, and individual identity. We even grant them loyalty, with consequences that John Marshall almost certainly never imagined.

Loyalty is surely a primitive virtue: our neolithic ancestors found it in the wolflike creatures that they domesticated as dogs, and then enhanced it by selective breeding. The German zoologist Helmut Hemmer claims that dogs, like all domesticated animals, have smaller brains than their wild forebears, but evidently loyalty and stupidity are perfectly compatible. Nondomesticated animals also show loyalty—parrots are loyal to their mates for life, mother alligators are loyal to their offspring for several years, and the honeybee workers will die for queen and hive. There are many stories of wild animals showing loyalty even to nonrelatives, in some cases to members of other species; some of these tales are probably true.

The point is that loyalty is not an invention of the Greeks, the Hebrews, or the Egyptians. Loyalty, or something very like it, is an ancient animal trait in which individuals defend or support other individuals despite a cost to themselves of energy, sustenance, or safety. But humans have expanded the scope of loyalty, applying it not only to each other, singly or in groups, and to members of other species, but to abstractions such as ideals and religions, and—here's the rub—to legal fictions: corporations and agencies that act when it suits their purposes as if they were real persons, albeit with immortality.

It is easy to see that loyalty in its original form has survival value, if not for oneself for one's genes. This applies to human tribal loyalties, loyalty to the place called home, and even to some national loyalties. The utilitarian advantage breaks down, however, when countries get so big and amorphous that they contain many conflicting loyalties, or when loyalties to abstractions become so extreme and ritualized that they threaten the welfare of the whole group and everyone around them. But to make a case against loyalties to nations and beliefs would also be to attack the driving force of much of civilization and its works. I don't want to do that.

What concerns me is a much more modern loyalty than the one to nations and beliefs, the loyalty to organizations. Should a person be loyal to an organization? Can an organization be loyal in return? Here I think we run up against a dangerous and self-destructive perversion of the ancient and honorable trait of loyalty. In law, an organization may be the perpetual embodiment of the goals and purposes of its founders; in real life, things are different. US Steel becomes USX. International Harvester becomes Navistar. The publisher Alfred A. Knopf sells out to Random House which had been purchased by RCA which in turn sells out to Advanced Publications, Inc., the Si Newhouse organization. The March of Dimes turns from a crusade against polio into a crusade against a motley assortment of afflictions lumped under the title of birth defects.

Imagine that you have a friend to whom you are fiercely loyal—call her Charlotte. One day you arrive at her house to find a note with your name on it pinned to the door: "Gone to Paraguay. Melissa will be your friend now." Silly? Transfers of loyalty that would be preposterous in the case of persons are pure routine when we deal with organizations. Not that changes in organizations are necessarily bad, but what does loyalty mean and what effects docs it have when one of the members of a loyalty relationship changes fundamentally? In a personal relationship such changes are often fatal to loyalty; in a person-organization relationship they are customary and we are asked to ignore them.

Sometimes there are no serious consequences when a corporate partner in a bond of loyalty undergoes a transformation. But if stability of character and purpose is a necessary part of a relationship involving loyalty, there can be trouble. We see this often in the business and academic worlds when a new team of administrators takes over, replacing or downgrading employees who have been loyal to the organization and who cannot understand why the loyalty is not reciprocated. We also see it when the care and protection of the environment are linked to long-term loyalty involving an organization. Like the dazed and disgruntled former employees who have been discarded, forests, lakes, and mountains may end up abandoned by the very organizations that once seemed pledged to protect them.

I am not referring to the wicked corporations that enlightened folks love to hate. Not D–w Chemical nor Ex–n nor even Gen-l E–c. The venerable villains I have in mind are loved by all, or at least by most.

Dripping with highly visible, avuncular solicitude like especially wise and benevolent Cheeryble brothers, these organizations can inspire blind, undiscriminating loyalty in all kinds of people ranging from starry-eyed innocents free of suspicion and doubt to shark-eyed robber barons not known for their open and trusting natures. One and all they are loyal to organizations that, if stripped of superficial appearances, they might hesitate to buy a used car from.

Take the case of the late Ernest G. Stillman, a wealthy epidemiologist from New York City. Dr. Stillman, who died in 1949, bequeathed the 3,800-acre Black Rock Forest near the Hudson River and West Point to his alma mater, Harvard University. He also gave a million-dollar endowment to care for the forest and to support the ongoing, long-term research on forest restoration for which its history and locality made it so ideally suited. Dr. Stillman had inherited most of the forest and much of his wealth from his father, a banker, but chose not to pass the forest on to any of his three sons. Perhaps he felt that it would be safer in the care of an immortal institution such as Harvard. Before Dr. Stillman's death, Harvard had agreed to accept the forest along with the endowment that would help protect it in perpetuity from the economic forces that had damaged it in the past and that Dr. Stillman knew would threaten its continued existence. In 1989, Harvard sold the Black Rock Forest, but kept most of the endowment, which by then amounted to more than three million dollars.

Harvard was quick to point out that there were mitigating factors. The forest was sold not to developers but to a conservation consortium put together during five years of negotiations by the philanthropist William T. Golden. The sale price of $400,000 plus $125,000 each from Harvard and Mr. Golden were combined to form a new albeit smaller endowment for the forest. The bulk of the original endowment was committed to support forest research at the Harvard Forest in Petersham, Massachusetts. Nevertheless, two of Mr. Stillman's sons were not happy with the sale of the forest and the transfer of the endowment. Both stopped their once generous contributions to the general fund at Harvard. "I'm saddened at the way I consider Harvard has breached faith with my father and ignored the trust they accepted forty years ago," John S. Stillman said to *The New York Times*.

But Daniel Steiner, vice president and general counsel of Harvard, stated, "We believe that what we are doing is consistent with the intent of Dr. Stillman's gift." Mr. Steiner is a successor to the Harvard corporation officers who had accepted the forest and the endowment from Dr. Stillman four decades earlier. One item that was not publicly discussed was how Harvard would define forest research and preservation at its remaining forest, in Petersham. Would Harvard, for example, use the endowment to pay for speculative and possibly controversial efforts of genetic engineers in laboratories in Cambridge who might be trying to create new, disease-resistant tree varieties? Would this be consistent with the intent of Dr. Stillman's gift?

Harvard is a corporation, immortal in American law since 1819; how can it have changed? George Trow, writing about Harvard and the Black Rock Forest in *The New Yorker* in 1984, explained: "What happened was that the work of running the university was separated from the work of maintaining its particular character." The generation that took over Harvard after the days of Dr. Ernest Stillman had a new view of what was important: they wanted Harvard to be world class in the modern sense, to transcend the merely local, even when the local could have wider application. The work of Harvard was now seen to be "everywhere and nowhere," to use once again the chilling phrase that Trow (and Mumford before him) used to describe contemporary rootlessness. Locality was for nostalgia, for fund-raising, for the patronizing smiles of tough, successful people who were being asked to give. The Black Rock Forest was quintessentially a place. Its sixty-year-old research projects were designed with that forest's history and particular ecology in mind. It was too local. It no longer fit Harvard's image. It had to go.

Around the time that Dr. Stillman was getting interested in forest research, Gifford Pinchot, America's first professional forester, was giving national recognition to the word "conservation" and was transforming the Department of Agriculture's Forestry Division into the U.S. Forest Service, which was established by Act of Congress in 1905. Pinchot's goal was to make the management of the national forests a showpiece of wise use consistent with perpetual preservation. Pinchot died in 1946, three years before Dr. Stillman, having turned over the reins of the Forest Service to successors a third of a century earlier. His name is still invoked as the guiding spirit of the Forest Service, but the Forest Service has changed.

The summer of 1989, which saw the sale of Harvard's Black Rock Forest, also saw the appearance of *Inner Voice,* an eight-page newspaper published by the fledgling Association of Forest Service Employees for Environmental Ethics and distributed both openly and clandestinely within the agency by its founder, Jeff DeBonis, a timber sale planner from Oregon's Willamette National Forest. Responding to the Forest Service's massive overcutting and overspraying of the federal forests, *Inner Voice* asked: ''Are you a frustrated Forest Service employee because your resource ethics conflict with your job?''

Putting his career at risk, the youthful DeBonis announced, ''We believe that the value system that presently dominates the Agency . . . is in need of immediate change.'' Two years later, AFSEEE, growing fast, had more than five thousand members, about half in government service and the rest private citizens. Jeff DeBonis had left the Forest Service and operated AFSEEE out of an office in Eugene, Oregon. *Inner Voice,* increased to sixteen pages, was circulated by its readers throughout the Forest Service and other resource agencies. It was mailed in unmarked wrappers and a notice on the last page stated, ''AFSEEE does not share its mailing lists with anyone. . . . Your name as a member is protected and completely confidential.'' Only by hiding their membership in AFSEEE could Forest Service employees be sure of avoiding the charge of disloyalty to the agency while promoting, however secretly, the reintroduction of the principles and standards of its founder, Gifford Pinchot.

The uses of loyalty are not as simple as they were in the days before the world was run by organizations. Loyalty to other people, the first human loyalty, is usually a wholesome trait. Loyalty to ideals, religious principles, communities, and nations can be similarly beneficial if exercised with care and judgment. Loyalty to organizations is a different matter, Chief Justice Marshall notwithstanding. By entrusting our wealth, leadership, and environments to the hoped-for unchanging individuality of immortal organizations, we subject our descendants to a likelihood of poverty, misdirection, and environmental degradation.

There are alternatives. When dealing with organizations, we must learn that loyalty is not an appropriate response. Less trust and more safeguards are in order. But there is another alternative, more basic, which we have never considered seriously since Thomas Jefferson first suggested it. In a letter to James Madison, in 1789, he said, "No society can make a perpetual constitution or even a perpetual law. . . . The earth belongs to the living and not to the dead." Jefferson believed that our most hallowed institutions and organizational structures should be scrapped and reinvented at regular intervals. This echoes a theme in Leviticus, where, in describing the provisions of the fiftieth, or jubilee, year, it was written: "But the land must not be sold in perpetuity; for the land is Mine; for you are strangers resident with Me."

Jefferson was worried about the tyranny of the ideas of the dead, but the perversion of the organizations the dead leave behind ought to be of equal concern. There is certainly danger in rejecting the organizations created by past generations—we may lose much of value. Yet the danger of being dominated by institutions whose purposes have changed or whose provisions do not meet an unanticipated present need is surely greater. We should regularly and unsentimentally check our institutions against the standards and goals of their founders. Nor should the original standards and goals themselves be above inspection. Loyalty to people and to ideals is admirable, but no institution deserves the kind of loyalty that is demanded by immortality.

Ecosystem Health

When I first became aware of environmental problems, in the 1960s, the environmentalists of the day commonly said that Lake Erie was dead. Even though it is difficult to imagine a lake as a corpse, the image was a vivid one and did much to arouse interest in pollution and other legitimate environmental concerns. Of course the lake was not dead, although its best friends were having trouble recognizing it. Perhaps it would have been better to say that the lake was in ill health.

The idea of health, which covers a spectrum of conditions ranging from being well to being dead, is applied to communities and ecosystems by both nonscientists and scientists. The new field of restoration ecology uses this terminology often, as we would expect because the concept of restoration itself is based on the idea of health. In the introduction to the book *Environmental Restoration* edited by John J. Berger, I found, for example, the following sentences: "In the Northeast, thousands of lakes are dead or near death, and forests, too, are sick and dying from acid rain and other pollution. . . . For the first time in human history, masses of people now realize not only that we must stop abusing the earth, but that we also must restore it to ecological health."

The image of a healthy environment is a compelling, natural, and probably ancient one, yet it also has origins in some of the ecological thought of this century. In the classic textbook *Principles of Animal*

Ecology, by Allee, Emerson, Park, Park, and Schmidt, published in 1949, a celebrated table compares seventeen supposedly shared properties of cells, multicellular organisms, and ecological communities. The properties include "anatomy," "regeneration of parts," "senescence and rejuvenescence," and "dynamic equilibrium." This is the organismic theory of ecology, the idea that communities are structurally and functionally like organisms; it is most closely associated with the name of Frederick E. Clements, who was active during the first four decades of the twentieth century.

Insofar as the organismic theory makes us compare communities to individual organisms, it is a normative interpretation of what communities are like. Your Aunt Margaret's parakeet is always a parakeet; it conforms more or less closely to some sort of platonic abstraction of what a parakeet should be: in its youth a lively, joyous parakeet, in middle age a calm, reflective parakeet, in old age a nasty senile parakeet, eventually a dead parakeet, but always a parakeet. You would be surprised if you looked in its cage and saw that it was turning into a lichen or a wombat. So it is with communities in the organismic view. They have a recognizable identity, and in the final stage of community embryology, or succession, that identity becomes fixed and normative: a prairie, a beech-sugar maple forest, a desert. Because communities have fixed identities, because they are normative like organisms, we can easily apply the normative idea of health to them: if they are functionally and structurally similar to their abstract ideal, they are healthy; if they deviate significantly, they are sick.

If the organismic theory of communities were still dominant, if the idea that communities have a normative, equilibrium position, a balance point, were still widely accepted, then the idea of ecological health would pose few problems. But concepts change in ecology, as in other fields of knowledge. Since the 1920s, but especially during the past two to three decades, the idea of the nature of communities and ecosystems has undergone revolutionary revision. No longer are communities considered normative by most ecologists.

The change of attitude was brought about because of awakening scientific interest in the phenomenon of natural environmental disturbance. Since the 1960s and 1970s, environmental disturbance has received increasing attention from ecologists. The reasons were initially partly pragmatic: disturbance and its effects are measureable; there are often funds available to study the economically important kinds of disturbance (for example, fire, invasion by gypsy moths or water hyacinths, changes in ground water level); and disturbance lends itself to the design of controlled experiments. Out of this preoccupation with disturbance has come an exciting new understanding of the workings of ecological systems along with the growing realization that natural disturbance is an essential part of the lives of many, perhaps most, ecological communities. The history of the last fifteen years of ecological research and thought has been a steady extension of our knowledge of this role of natural disturbance in the life of communities and a corresponding erosion of the normative view of community organization and function.

Among the kinds of communities and ecosystems now known to be frequently influenced by disturbance are temperate deciduous forests, many coniferous forests including boreal and coastal plain forests, tropical evergreen forests, grasslands, and some aquatic systems. Even deserts and tundra are affected by natural disturbance. Disturbances occur both spatially and temporally; in the latter case, disturbance exists on time scales measured in hours, days, months, years, and millennia, depending, for example, on whether one is looking at tides, fires, virus cycles, or climatic change as agents. Communities and ecosystems are in constant flux, even without human interference. How, then, do we define health, which depends on a definition of a normal state, and how do we separate, except in the most obvious cases, the effects of human and natural disturbances? As Steward Pickett and Peter White stated, in the summary chapter of their edited book entitled *Natural Disturbance and Patch Dynamics,* "An essential paradox of wilderness conservation is that we seek to preserve what must change." Equally paradoxical, in some cases like the Kansas prairie, we must work to preserve disturbance, such as frequent fires, in order to preserve what we have come to think of as the essential character of particular ecosystems.

The discovery by ecologists of disturbance has caught the public eye. On July 31, 1990, the lead article in the science section of *The New York Times* was headlined "New Eye on Nature: The Real Constant is Eternal Turmoil." The writer, William K. Stevens, was alert to the environmental implications of disturbance theory, partic-

ularly of the discrediting of the organismic notion that there is a
balance of nature, a healthy equilibrium state in communities that
have not been unduly affected by human actions. Although the article was careful to point out that even disturbance ecologists admit to
"a kind of equilibrium . . . on some scales of time and space," the
subheadlines conveyed no such hesitation: "A Difficulty—Posing a
Question: What is Natural?" and "Empty Theory—Observations
Find No Neat Balance."

Before examining the implications of this philosophical shift for
the concept of ecological health, I want briefly to look at the change
in ecological theory in a different way. Natural disturbance has always existed in communities, and ecologists have known about it for
a very long time. But prior to the 1970s the prevailing tendency was
to minimize it in favor of data and interpretations that supported the
idea of homeostatic, organismic balance. Why? I think it was because these ecologists were themselves part of a human environment
that instilled a strong, highly developed sense of a normative community, a balance. In the United States, at the time when the organismic theory developed, there was a powerful sense of what constituted the equilibrium position of this national community. This sense
of a national equilibrium was not fundamentally challenged by either
the Great Depression or the Second World War, both of which disturbances were seen as sicknesses or departures from a normal condition and were treated accordingly without overwhelming feelings of
hesitation or uncertainty. It is not surprising that ecologists during
this period saw nature as made up of definable, balance-maintaining
communities, much in the same way they saw their own country.

Then came the military-industrial complex, the Vietnam War, the
crisis of urban life, the clash of materialism and idealism, the status
quo versus the women's movement and the struggle for racial equal-
ity, upheaval in the universities, speciesism versus animal rights, and
the decline of communal morality and the apotheosis of the individual. Where was the balance point? Is it a coincidence that the individualistic theory of community structure and the changing view of
stability and diversity began to be widely accepted around the time
that the anti-establishment movie "The Graduate" was first shown?
Is it a coincidence that mathematical ecologists had a brief fling with
catastrophe theory in the late seventies, and later, as the Soviet Union

disintegrated and the United States, Israel, and Syria jointly opposed Iraq, became preoccupied with chaos? I do not think these are coincidences. For twenty years, the idea of a national norm, an equilibrium position, had been breaking up under the influence of repeated disturbances. Forty-five years ago, W. Clyde Allee could write professionally about the organismic theory and "protocooperation" within animal populations, while in his private life advocating world unification and peace. I don't know what his theoretical position would be if he were alive in these times.

It would be naive to assume that the nonequilibrium theories of community and ecosystem structure are likely to be the last word on the subject. But to observe that ecological theory appears to be culturally dependent and changeable is not to invalidate the present form of that theory. The modern perspective raises real and serious problems for the concept of ecosystem health. First, as I have already said, it is extremely difficult to determine a normal state for communities whose measurable properties are often in a condition of flux because of natural disturbance. David Schindler, a Canadian ecologist, in a careful review of aquatic ecosystem responses to human-caused stress, said, "The fact is that we usually do not know the normal range for any variable, at least for any time period greater than a few years. . . . Even if well-designed monitoring programs were magically emplaced tomorrow, it would be years before we could confidently distinguish between natural variation and low-level effects of perturbations on ecosystems."

Of course human-caused disturbances can be orders of magnitude greater than natural ones (or faster, if we are referring to disturbances such as climatic change); this can help establish at least a comparative equilibrium for the purposes of defining health. Or disturbance and major fluctuations at one scale of observation, for example the population scale, may not be matched by fluctuations at a different scale, for example the scale that integrates many populations—the community. Also, Schindler noted that there were various ways of getting around the problem of defining a healthy baseline in the face of normal variation. One of these ways is to deduce the species composition of plant communities of prehuman eras by identifying and counting the grains of fossil pollen from the soil deposits of those eras. Another is to disturb whole ecosystems, such as small lakes, in a carefully controlled experiment and measure the biological and physical changes that take place.

To confuse matters more, equilibrium theory is not really dead. Some ecologists have shown that despite the prevailing fad for non-equilibrium theories in ecology, many communities still behave in a way that more or less suggests equilibrium. Often, the decision of whether a community is fluctuating in a normative or nonnormative way is a subjective one, a matter of deciding whether numbers fall on one side or another of an imaginary line drawn by the researcher. Where the line is put may well be influenced by subconscious bias for or against the notion of a normative, balanced equilibrium in nature. Nevertheless, the problem of nonnormative, nonequilibrium systems has been raised and will not go away.

Another problem with the idea of ecosystem health is that ecosystems have many functions and processes, not all of them strongly related to each other. A judgment of ecosystem health can be a function of which process you are looking at, which in turn is determined by your own values. The accident at Three Mile Island may have been made much more serious by control room operators who had no way of knowing which of the many kinds of reactor variables to pay most attention to during the emergency—reactor water pressure, water temperature, water level, etc. Some readings remained normal; others did not. Of course the same issue could be raised about any judgment of human health, which is highly dependent on which variables are observed, yet we don't discard the idea of health for that reason. But again, the problem exists.

Third, a general word such as health can end up with all kinds of narrowing qualifications and can lose some of its original meaning if we apply it too rigorously to examples of specific communities. Because communities vary greatly, with some occurring at the equilibrium end of the range, some at the nonequilibrium end, and others at all degrees in between, the larger, unifying idea of health can vanish as we redefine it from case to case. Any attempt to imbue ecological health with the rigor and specificity that will allow us to use it as a scientific tool may well strip it of the intuitive, general meaning that is its chief value. We have an example of this in medicine, where a five-year remission from cancer following strenuous treatment of the disease may be equated with health by some medical statisticians, but not necessarily by the patients whose definition of health embraces a far larger world of experience.

In other words, health is a judgment that can be made only by someone who has been intensely familiar for a long time with what is being judged. Each living system is unique and includes too many variables to let us expect that the mere application of scientific formulas will do the job. In medicine, the best judge of health is usually the patient, who has had the longest and most intimate acquaintance with the body and mind that are being evaluated. This is why doctors who don't listen to their patients are so incompetent and so rightly despised. Who, then, is the best judge of the health of an ecosystem? Clearly the one who knows that ecosystem best—by virtue of close observation over many years. This person may be an ecologist, but equally well may be an amateur naturalist or a resident of the area, possibly a resident without any scientific training. There is no technical expertise that automatically confers the ability (or right) to judge ecosystem health.

Scientists thus have no exclusive right to proclaim ecosystems healthy or sick. Reluctance to share the term *health* with laymen coupled with the scientific obstacles to using the word in ecology has made some ecologists shy away from the whole notion of ecosystem health. That is too bad. By being unwilling to come to grips with the concept of health, they sometimes give the appearance of sanctioning the destruction of ecosystems.

Ecologists should not be afraid to use the term *health* at least occasionally, even though it may have no place in their daily research. They need not define it in most cases; when it is defined it should be explained in a durable, general way that is as independent as possible of particular ecological theories. Such definitions may involve the rates at which changes in ecosystems have occurred, in addition to the usual considerations of loss (or gain) of species and changes in physical structure and chemical properties. They will always involve informed but unmeasurable human perceptions of what is happening to the ecosystem.

Health is an idea that transcends scientific definition. It contains values, which are not amenable to scientific methods of exploration but are no less important or necessary because of that. Health is a bridging concept connecting two worlds—it is not operational within science if you try to pin it down, yet it can enable scientists and nonscientists to communicate with each other. Equally important is its effect on the scientists themselves. If used with care within ecology, the idea of health can enrich scientific thought with the values and judgments that make science a useful and durable human endeavor.

Down from the Pedestal— A New Role for Experts

We live in a society that worships expert knowledge, and sometimes worships the experts themselves. This can be very pleasant if you're an expert, but has several curious and important drawbacks for the rest of us.

The first and maybe the most important drawback is that expert knowledge is rarely sufficient for analysis, prediction, and management of a given situation, with the exception of basic repair operations. This is because in order to limit the number of variables they have to contend with, experts make the assumption that the systems they are working with are self-contained, defined, closed. Real-life systems are hardly ever closed.

Naturally, some systems conform better to the assumptions of the experts than others; they are closer to being closed. So there is a range of success which is as much dependent on the field as it is on the expert him- or herself. At the dismal end of the range are those open-system fields where expert knowledge, especially concerning long-range prediction and design, is worth less than its face value, and is sometimes not much more useful in making sound decisions than a random number table. In these fields, expertise can be a pose, heavily dependent on professional jargon and a smokescreen of mathematics, statistics, and technology. The pose usually works.

Some obvious examples are economics, long-term weather forecasting, and educational psychology.

At the other end of the range, where the expert's assumptions of a closed system can approximate reality, are such fields as routine dentistry and air traffic control. Some of us might not want to be dentists or air traffic controllers, but we have to admit that, given a reasonable degree of professional competence, they usually get the job done as specified in advance.

Where do the expert fields of ecology and wildlife management fit in this range? Certainly they are not as bad as economics or weather forecasting; there is more ability to make pretty good, long-term, or at least mid-term, predictions that show trends in the right direction and within an order of magnitude of the actual numbers. On the other hand, the science and management of natural resources are a far cry from dentistry or television repair in the correspondence between predictions and actual events or in the fulfillment of planned objectives. In his celebrated, 1976 keynote address to the annual meeting of the American Fisheries Society, in Michigan—an address entitled "An epitaph for the concept of maximum sustained yield" and directed towards an audience of expert fisheries biologists—Philip A. Larkin began to define the problem:

> Natural systems are sufficiently diverse and complex that there is no single, simple recipe for harvesting that can be applied universally. When there is added in the complexity and variety of social, economic, and political systems, the number of potential recipes is just too enormous to be easily summarized by simple dogma. Perhaps the best we can hope for is a general statement of principles with accompanying guidelines that should be applied in the hope of ensuring that we will trend in the best direction.

Although Larkin's paper was about a specific management issue, I think that the quotation has general usefulness. The reductionist methods of science, which can work extremely well in closed systems, tend to break down under the open-endedness imposed by biological complexity and by the interacting complexities of political, economic, and social factors.

Disappearing Salmon and Other Surprises

We can take biological complexity first. Despite what is sometimes described as the march of progress, I don't think there has been any significant change since Larkin's address in 1976. In a paper published in the January 1991 issue of *BioScience,* C. N. Spencer, B. R. McClelland and J. A. Stanford described the interesting and unexpected chain of events that followed the deliberate introduction of opossum shrimp, which salmon love to eat, into more than a hundred lakes in northwestern North America. Originally intended to stimulate the production of kokanee salmon, the opossum shrimp appear to have had the exactly opposite effect, at least in Flathead Lake, where Spencer et al. studied them. The story is a complicated one, with nutrient loads, water levels, algae, various invertebrates, and lake trout all interacting. But the bottom line is that the kokanee salmon population went way down rather than way up, and this in turn affected populations of bald eagles, various species of gulls and ducks, coyotes, minks, river otters, grizzly bears, and human visitors to Glacier National Park.

If the system had been limited to the salmon and the shrimp, things might have gone as predicted, although even that isn't clear because the shrimp and the salmon normally occupy different parts of the water column during the daylight hours, which is when the salmon can see to feed on the shrimp. At night, when the shrimp rise to the upper levels to feed on zooplankton, they are safe from the salmon. At any rate, the system was not limited to the shrimp and the salmon. I suppose a fairly predictive model could be constructed now— retrospectively predictive, that is, of the events that have already been documented. I doubt that it could have been constructed before the fact, however, or, if it had been, that it would have necessarily prevailed over the models that predicted the opposite effect.

The biological complexities that I have described so far concern interactions between species. Within species, however, the biological imponderables may be as great. This is true not just for obscure organisms, but also for much-studied, commercially important species. To look at another example from fisheries biology, take the case of Pacific salmon, described in a paper in *Conservation Biology* in

1990 by R. S. Waples and D. J. Teel. Even in confined and regulated hatchery stocks, as Waples and Teel, and others, have shown, it is hard to tell whether observed genetic changes are because of deliberate selection or intrinsic genetic drift; it is hard to tell how many effective breeding females there are; it is hard, maybe impossible, to predict the long-term effects, if any, of observed genetic changes; and it is hard, maybe impossible, to predict the ultimate consequences of the mixing—accidental or deliberate—of hatchery and wild stocks of any given species. So the complexity and uncertainty of biological systems go far beyond food webs: they are everywhere we look. This means that biological complexity, with its myriad internal and external variables, with its open-endedness, pushes ecology and wildlife management a little closer to the economics than to the dentistry end of the range of expert reliability.

When the nonbiological variables are factored in, all pretense at a closed, definable system crumbles. Take the case of nutrient loading, one of the factors that appears to have influenced the abundance of kokanee salmon in various lakes. Nutrient loading, the input of nitrogen, phosphorus, and other nutrients into lakes, streams, and bays, is itself determined by scores of human activities, including sewerage and other household discharges, agricultural practices, air pollution, especially with nitrates and nitrites, industrial point source discharges, and others. Each of these is strongly influenced by many variables: for example, sewerage and household discharges depend on cultural practices and on the economics, politics, and technology of sewerage treatment, which is in turn determined by an interplay of forces at the local, regional, and national levels.

Of course one can bypass all this uncertainty by priming a model with ranges of numbers for any given variable rather than with a specific figure, but when this is done for several variables, the model quickly changes from a forecast to something altogether less pretentious, albeit still useful, something akin to the cautious economist's or planner's feasibility study. E. F. Schumacher, a cautious and very realistic economist, wrote: "A long term forecast . . . is presumptuous; but a long-term feasibility study is a piece of humble and unpretentious work which we shall neglect at our peril. . . . It is surely a sign of statistical illiteracy to confuse the two." A feasibility study as Schumacher used the term is very much like those wildlife and fisheries projections that extrapolate trends based on ranges of variables, and which come up not with specific predictions, but with highly conditional ranges of answers. In this there is something both a little discouraging and a little hopeful; I will come back to it shortly.

I have tried to show that when experts apply the methods of objective science in an area like fisheries management—let alone educational psychology—the results are not as clear-cut as the attempts of astronomers to predict the orbits of moons and asteroids, or of engineers to calculate the effects of stresses on a bridge girder, or of orthopedic surgeons to estimate the number of centimeters of bone growth remaining for a twelve-year-old girl. But to the lay public, an expert is an expert: once you put on a lab coat or sit down at a computer, once you get a graduate degree, carry an attaché case, publish papers, or use tax moneys for research, you are an expert and you have the answers. It is no use getting angry at the public for this attitude; they will only think you are not a real expert after all.

The Horoscope Method of Forecasting

It seems to be human nature to have exaggerated respect for the abilities of people who have special, inaccessible knowledge. The more specialized and arcane the knowledge, and the more it confers power over life and matter, the greater the awe of the layman. Formerly, and occasionally still, this awe was reserved for magicians, shamans, and astrologers—now it is directed at nuclear physicists, high-tech doctors, and even marine scientists. And we experts respond by fostering this aura of magic and power, partly because we believe it, partly because it's fun, and partly because it helps us get jobs and grants. I know that ecologists and wildlife biologists are not in the same league for hubris as nuclear engineers and transplant surgeons, but experts are experts, and few escape the lure of being put on a pedestal.

So now we are stuck on the pedestal. We know much, much more about our subject than the average citizen does, but much, much less than the average citizen expects and demands us to know. The public knows that we each have a crystal ball—we know that the ball is opaque; when we look into it all we see is a distorted reflection of ourselves and everything around us. How can we cope with this difficult situation?

There are only two ways that most experts have found to deal with the problem. They are mutually exclusive, and I do not think that either of them is acceptable. Before I describe them, I should say that I am certainly not singling out ecologists here—all experts are pretty much in the same leaky boat. The first way of coping with the public's excessive expectations is what I call the Horoscope Method. The Horoscope Method is fundamentally dishonest, but that does not bother its practitioners. In this method, we experts assume that the public is right, that we do have the answers. We believe implicitly in our models. The more specific their predictions are, the more we believe in them, no matter how scientifically preposterous and absurd that specificity is.

The reliability of long-term forecasts tends to be inversely related to their specificity and usefulness. Every horoscope writer knows that if you stay general, you're safe. Here is the horoscope for somebody born today. I know it is accurate because I made it up very carefully myself. "You are worried about money today. Take the advice of a relative or friend. This is not the time to assert yourself." But when was the last time you read a horoscope that said, "You are going to develop a left inguinal hernia today. A check you wrote for $106.14 will be returned for insufficient funds"? No writer of horoscopes would dare to get so specific. The experts, however, who are generally a more naive and self-confident bunch than the horoscope writers, are not so cynically cautious. I am reminded of some re-markably specific—and totally wrong—earthquake forecasts for Missouri that put many people in a state of panic.

The Dignified Retreat

The other method of dealing with the public's inflated expectations is what I refer to as the Dignified Retreat Method. This is far more popular than the Horoscope Method because it does not require a masterful personality to put it across and because it is not overtly dishonest. There are two variants of this method which shade into one another. First is the attempt to confine one's activities and statements to the strict scientific or technical subject matter of one's expertise. "Don't talk to me about politics and economics, I have to examine these microscope slides." Or, "If you can't relate your question to the numbers on this computer printout, I hardly see how I can be expected to answer it." These experts include the ones who assert the superiority of pure over applied science, although the school of knowledge for the sake of knowledge has been declining ever since the genetic engineers started to get salaries over one hundred thousand dollars a year. This version of the Dignified Retreat Method considers off limits any question whose origins are broader than one's specialty. It is the product of a fortress mentality: what goes on outside the fortress is not important.

In the war room of the journal I edit, *Conservation Biology*, I am beginning to receive reports of scattered sniping from colleagues inside the fortress, angry because I print other things in the journal besides refereed papers with numbers in them. For example, there are opinion columns by David Orr, chairman of environmental studies at Oberlin College, who suggests that it is dangerous for conservation biologists to assume that professional success in their scientific research is necessarily correlated with any improvement in "the health of the biotic world." "As important as research is," Orr writes, "the lack of it is not the limiting factor in the conservation of biological diversity." I do not think that much conservation science and management are likely to have lasting environmental benefit unless such subversive ideas are understood by the experts doing the work.

The other variant of the Dignified Retreat Method is a little less insular but not much better at answering the public's questions or meeting its needs for understanding what is happening to the environment and what our conservation options are. At *Conservation Biology*, I often get papers from the practitioners of this variant. They are all the same. After twenty pages of highly technical data and discussion, some of it incomprehensible or meaningless, there is a brief concluding paragraph which starts, "These findings have important implications for conservation," and then goes on to express a conclusion that all but the most seriously retarded third graders would find self-evident. The last sentence of the paragraph is inevitably a call for more research. True, the writers of these papers have stepped outside the fortress to confront the public, but the meeting is so brief and they are so heavily shielded by their science that they might as well have stayed behind the walls.

Honest, Effective Experts

Is it possible for experts to cope with the uncertainty and open-endedness of their work in the face of unreasonable public expectations? Is there a better alternative to the Horoscope Method or the Dignified Retreat? I think there is, but it is both difficult to practice and occasionally painful. Difficult, because it means reconciling two very different world views so that they can be employed simultaneously; painful, because it means giving up the special aura and mystique that sets the expert apart from the lay person and provides a comfortable feeling of prestige and power.

One of the first to demonstrate this third alternative, who showed us how experts could be honest, effective, and communicative without either losing the respect of the public or abandoning professional competence, was E. F. Schumacher. Rachel Carson was another. Schumacher was an expert, considered one of the most brilliant economists of his time, and chief economist of the British Coal Board, then the largest socialized industry in the West. Although not a forecaster, he was one of the first to predict the rise of OPEC and to assess its implications. Schumacher recognized the limitations of expertise in economics, but instead of denying them or hiding behind techniques and procedures, he transcended the limitations.

"Economics, and even more so applied economics," he wrote in *Small Is Beautiful,* "is not an exact science; it is in fact, or ought to be, something much greater: a branch of wisdom." Much of his later work was aimed at just that, making economics a branch of wisdom. Of his many accomplishments toward this end, I will cite only two. His writings on Buddhist economics used religious ideas that were previously completely alien to economics, from another world view, to begin to transform the economic theories related to labor. That this transformation is still only beginning is an indication of the inertia and fear of change inherent in economics and many other expert fields. And Schumacher created the idea of intermediate technology, at the core of which is the belief that experts can apply their special knowledge in cooperation with the public in an open, fully interactive, and mutually respectful and beneficial way.

More recently, Herman E. Daly and John B. Cobb, Jr., the former a senior economist at the World Bank and the latter a professor of philosophy and theology at Claremont College, in California, have continued in this tradition. In their book *For the Common Good,* they began with a chapter entitled "The fallacy of misplaced concreteness in economics and other disciplines," and ended with a chapter entitled "The religious vision." In the first chapter, they quoted from a letter published in *Science* by Wassily Leontieff, a mathematical economist and Nobel laureate:

> Page after page of professional economic journals are filled with mathematics formulas leading the reader from sets of more or less plausible but entirely arbitrary assumptions to precisely stated but irrelevant theoretical conclusions. . . . Econometricians fit algebraic functions of all possible shapes to essentially the same sets of data without being able to advance, in any perceptible way, a systematic understanding of a real economic system.

I said at the beginning that ecologists are not as bad at dealing with the world as economists, but that does not mean that we have nothing to learn from the example of economics. What Daly and Cobb, if not Leontieff, are saying is that for experts dealing with real-world problems, the world view of professional, technical expertise is not enough, except possibly for dentists, air traffic controllers, and others whose professional world is relatively circumscribed.

So that I am not misunderstood, I have to add a parenthetic comment. Claiming an ability to predict the future based on one's models is not the same as airing rational and reasonable fears that are based on special knowledge and experience. Scientists have an obligation to speak out in their fields if their speech has a good chance of helping the public avoid doing serious environmental damage. The ecologist F. Herbert Bormann said, at the Ecological Society meeting in Fort Collins, Colorado, in 1984, that it was important for biologists to warn laymen about acid rain even before definitive answers were available. This kind of warning, based on a large body of circumstantial evidence presented with care, qualified, and subject to discussion, is a service, not a threat, to the public.

Talking—and Listening—to the Public

The task at hand is not just one of spiritual development or expansion of world view on the part of the experts. I think society will also find it necessary to sacrifice its long-cherished idea of the expert as someone separate, gaining status by being apart. In a practical sense, what this means is more education about what experts do. Every grant and every project with an environmental component should make provision, I mean funding, for a component of public education about that work. Why should people value what experts do if they do not know what it is or what it is for? How can experts claim to be serving the taxpayers if they never talk to them about that service? If you work for somebody and avoid ever seeing that person, then there is something wrong.

The education that has to be built into the regular work of experts can range in level from kindergarten to adult. It should go on during the life of the grant or project, not just when it is over, so that any useful suggestions from the public can be used to modify the work in progress, before it is too late. Experts can learn, too. Funds could come in part from the bloated research overheads now being charged by private institutions or indirectly extracted by government agency bureaucracies—if a few surplus administrators were forced to seek meaningful employment elsewhere, that would be an added benefit. Education is a skill: learning how to teach effectively about their technical work, and how to listen, will not be easy, but people smart enough to become experts ought to be able to pick it up.

These changes are essential and there is not much time to bring them about. It is a matter of environmental survival *and,* ultimately, the survival of publicly supported experts in a time of declining national wealth and resources. The environment has not been well served by experts in isolation, or by experts who pander to the public's exaggerated ideas of their powers. Abandoning these roles in favor of a more responsible alternative, difficult as that may be, will in the long run be the least painful choice, both for the experts and for the environment.

Asking Unaskable Questions—
A Case Study

There aren't many people whose mere thoughts are more interesting than their way of making money, their private sex lives, the story of their miraculous recovery, their method for training police dogs or losing weight, or their six gourmet recipes for leftover broccoli. Garrett Hardin, professor emeritus of human ecology at the University of California at Santa Barbara, is one of the precious few: the original thinker's original thinker.

So forget the usual stuff of biography: the peripatetic life as a younger son of a father employed by the Illinois Central Railroad; the lonely, exciting days on his grandfather's farm in Missouri; the polio that left him, at the age of four, partially crippled for life; the swimming; the wife and four children; even the details of his pro-abortion battles of the 1960s. What matters are Hardin's ideas, the shimmering, crystalline creations of his mind that never fail to astonish his admirers and that leave his many enemies feeling as if they were trying to ward off a giant with a willow twig.

In the story of the emperor's new clothes, the emperor's nakedness was finally exposed by a little child who lacked the sophistication to rationalize away what he could plainly see. Hardin's role in life has been that of a child pointing at naked emperors (the title of one of his books of essays is *Naked Emperors*). He is a hunter of taboos, of the concepts and practices we declare sacred, out of bounds to criticism and rational discussion. And like anyone who questions the sacred, he has aroused more than his share of opposition from all quarters.

Both liberals and conservatives can find plenty to make them nervous in Hardin's writings. In fact, if you can read Hardin for more than ten minutes without getting mad, you are probably (1) a saint, (2) brain-damaged, or (3) a very slow reader. Liberals he dismisses, using Michael Novak's explanation, as people with no human "middle term" between individualism and universalism. Conservatives he embarrasses by rejecting such contemporary right-wing causes as "right-to-life," efficiency, and perpetual growth. Truly, the professor has never worried about his public relations image.

The article on earthquake prediction in his book of essays called *Stalking the Wild Taboo* is a case in point. This essay, he writes proudly by way of introduction, was rejected by more magazines than anything else he has written. And no wonder. In it Hardin makes the case for not spending any federal money on new research in earthquake prediction. (Note that he lives in Santa Barbara; at least he can't be accused of cowardice.) "At best," Hardin says in this essay, "if the research were unsuccessful, $137 million would be wasted." This is the best that could come of such a scientific endeavor. The worst? You guessed it: the research might succeed and we might learn how to predict earthquakes.

There is a fundamental difference, Hardin observes, between earthquakes and hurricanes. Hurricanes already exist when we start to predict their behavior. Storm forecasts give us at most a day or two in which to take fright and do whatever we are going to do—batten down or flee. But earthquakes are another matter. "On the basis of all we know about earthquakes, it is almost certain that in an earthquake warning system, lead-time and reliability would be *directly* related." In other words, if scientists learn how to predict earthquakes reliably, the chances are that such predictions would be made six months rather than six days before the actual quake. (It appears that the Chinese, years after Hardin wrote about earthquake prediction, were able to evacuate one city shortly before a major quake, with a resulting saving of life. But this kind of last-minute warning is not the sort of forecasting system at issue here.) So assume that scientists do learn to predict earthquakes several months before they happen. "And then what?" he asks.

All hell would break loose. What could anyone do, knowing that a quake was probably coming during the next few months? Pack the china away? "What do you eat off of in the meantime?" Hardin wonders. Think of the psychological effects on a population already enduring the multiple stresses of life in the fast lane. Think of what would happen to real estate values and home sales, to the insurance industry, to banking, to stock values of local industries, to construction. . . . At the first hint of a forthcoming prediction of a quake, Hardin continues, powerful interests would inevitably combine to suppress it. This effort would be followed by the equally inevitable rumors, leaks, and profiteering, all adding to the social turmoil. Some people would be leaving the earthquake zone, some would be staying, others—experts at exploiting chaos and human misery—would be arriving in unsavory droves. Could the quake itself, when it finally came, possibly cause as much damage to society as its prediction?

What bothers Hardin most is that the federal agencies that have considered whether to fund earthquake research have never even thought about the side effects of prediction.

> Nowhere, do they indicate that they have the slightest inkling of the psychological and sociological dangers of the scientific advance they hope for. Their analysis is a beautiful example of the pathology of specialism. . . . In their bland indifference to the total meaning of earthquakes these geophysicists remind us of the apocryphal surgeon who 'solved' the problem of the weeping patient by extirpating her tear glands.

The method of Garrett Hardin's inquiry is organized around one critical question: "And then what?" This is the ecologist's question par excellence. Ecology, the study of living organisms in their environments, is above all concerned with interactions and their consequences. What Hardin has done is turn an ecologist's eye on processes and happenings that most of us take for granted, and to look beyond at the consequences. He likes nothing better than to take our society's most beloved ideals, the ones we have painted a gleaming white, and apply a few coats of intellectual paint remover. Even the titles of his essays reveal his method. "In Praise of Waste," "The Moral Threat of Personal Medicine," "Vulnerability—The Strength of Science," "The Threat of Clarity," "Nobody Ever Dies of Over-Population." Do such themes seem paradoxical to you? You didn't look far enough, says Hardin. Ask the question "And then what?" and the other person's paradox will become your common sense.

Hardin is a taboo-hunter, and such game is plentiful, but he doesn't scatter his shots at random. His principal target—pursued with a near-obsessional zeal for more than a quarter of a century—is the democratic and liberal notion that what is good and true for the individual must by extension be good and true for everyone. His great assault on this doctrine and certainly his best-known work was an article originally published in the journal *Science,* in 1968, and since then reprinted in many places. He called it "The Tragedy of the Commons." At the heart of this profound and indeed tragic essay is, as we might expect from Hardin, a paradox—a deceptively simple one. Consider a "commons" such as the original commons or public grazing land. Any herdsman has the right to graze cows there, as many as desired. Assume that the commons is already crowded, and you, a herdsman, want to add another cow to your herd. What is the benefit to you? Clearly one-cow's-worth of benefit, because all the profits will come to you. And what are the costs? Because grazing is free, the ongoing costs are primarily those related to overgrazing. But the one-cow's-worth of overgrazing costs contributed by your extra cow is shared by everyone who uses the commons, and you will only perceive a tiny fraction of it. The profits are yours but the costs are everyone's. With this kind of accounting, why not buy another cow? And then another. That's the tragedy of the commons: what's good for you alone may be awful for you plus your neighbors.

Although Hardin has identified deteriorating commons in many areas of modern life, from medicine to pollution, his bête noire, the commons that frightens him most, is the right to multiply, with its result not of overgrazing but of human overpopulation. In "Nobody Ever Dies of Overpopulation," a short essay written in 1971, he made clear his belief that the world is already overpopulated. De-scribing a storm that had just killed a half-million people in East Bengal (now Bangladesh)—people who had been forced by overcrowding into the low-lying, unsuitable land of the Ganges delta—he asked:

> What killed those unfortunate people? The cyclone, newspapers said. But one can just as logically say that overpopulation killed them. . . . In the web of life every event has many antecedents. Only by an arbitrary decision can we designate a single antecedent as "cause." . . . Were we to identify overpopulation as the cause of a half-million deaths, we would threaten ourselves with a question to which we do not know the answer: How can we control population without recourse to repugnant measures? . . . Instead we say that a cyclone caused the deaths. . . . Fate is so comforting.

How can we control population, given the tragedy of the commons? Enclose the commons, says Hardin, restrict the right to multiply. Keep our distance from other places that have refused to enclose their commons. "Mutual coercion mutually agreed upon" is his phrase. Limit some personal freedoms in order to preserve others more precious.

And how do we—even a simple majority "we"—mutually agree upon and bring about such mutual coercion? On that subject he is less clear. Garrett Hardin is better at exposing the nakedness of foolish emperors than he is at sewing new clothes for them, which may turn some away from his writings. But it should not be held against him. An overpopulated world needs clear-sighted children who dare to ask unaskable questions. The answers will follow.

Changing the Way
We Farm

Mountain Sheep and Moose

In 1971, the Canadian wildlife biologist Valerius Geist published his book, *Mountain Sheep: A Study in Behavior and Evolution*. In it he observed that although both the United States and what was then the Soviet Union had enacted measures to protect mountain sheep or to control hunting in areas where they were still common, and although vast areas of suitable sheep habitat, without sheep, still exist, mountain sheep have made "no noteworthy recovery on either continent. Protection of sheep has been locally in force in the United States since the turn of the century and in Russia since after the October Revolution, yet only exceptionally have sheep spread from the restricted areas to which they were confined." This contrasts sharply with the fate of moose, which have taken advantage of twentieth-century protection to recolonize thousands of square miles of terrain in Asia, Europe, and North America that they once occupied and from which they were exterminated. Why the difference, and what is its significance for farming?

Geist gives an ecological and evolutionary answer to the question of why mountain sheep are so reluctant to reoccupy suitable former habitat that is nearby. The explanation has little to do with physical barriers. No terrestrial mammals, including moose, can equal the mountain caprids—sheep, goats, and ibexes—at crossing long stretches of rugged landscape. The explanation is more subtle: moose are creatures of the fluctuating and transient lowland habitats that follow forest fires and floods. Rarely can they stay in one place for more than a few years because the vegetation that follows fires and floods—the plants that they eat—is rapidly replaced by less suitable kinds of trees and shrubs. Not surprisingly, they have evolved a behavior that adapts them to explore and colonize new areas when they are still yearlings.

Mountain sheep, by contrast, occupy highly stable and long-lasting grass communities that persist from decade to decade. This habitat is patchy in distribution and the patches are separated by large stretches of unsuitable wooded and other habitats. Also, summer and winter ranges may be miles apart. Within the territory of a particular band of sheep, all suitable patches are known and used, and, as Geist learned, this information is culturally transmitted from the older females who lead the band to younger sheep in successive generations. Almost without exception, young mountain sheep follow older sheep. They do not explore. The risks are too great, and the few with a tendency to stray are not likely to survive long. In mountain sheep, the natural extension of range occurs on a geological scale of time, as a consequence of extremely rare, chance events. Now set the mountain sheep aside for a while; I will come back to them.

Agriculture in Transition

The agricultural system known as the Green Revolution which was developed in the 1950s and 1960s, which expanded in the 1970s, and which began to fall apart in the 1980s, needs to be replaced. Based on the intensive use of fertilizer, irrigation, insecticides, herbicides, fungicides, farm machinery, and large, single-crop plantings—all in one take-it-or-leave-it package—the system has defects that are major and inescapable. It costs a fortune, making farmers mortgage and risk their assets every year. It leaves farmers no control over the way they farm and little control over the way they market their produce. The chemicals that are an inseparable part of the system are highly toxic to the farmers and their environments. It erodes and wastes the soil even as it poisons it. It depletes scarce water resources. It allows, even promotes, the extinction of countless precious varieties of crops, the irreplaceable genetic heritage of millennia of farming. And all these consequences, acting together, have destroyed farm culture and farm communities and have forced millions of knowledgeable farmers to abandon farming and leave their land, in the rich and poor nations alike.

So much is clear. Now the hard part: what will replace the Green Revolution? We have a spectrum of choices ranging from conservative tinkering that redesigns but does not alter the underlying system to radical change that overthrows it completely. In developing the next agriculture there is no latitude for error. We have squandered our safety margin by ruining our land and the farm culture that used it. Conservative tinkering may not address the basic problems; radical restructuring may not work at all. This is the unvarnished prospect.

At the conservative end of the spectrum are many techniques and options, the majority of them not ecological in nature. Efficient, Israeli-style irrigation systems. Creatively financed farmers' cooperatives to process the food where it is grown. New marketing strategies including direct farmer-to-consumer sales and u-pick-it operations. An end to crop price subsidies, which are in effect subsidies to manufacturers of farm chemicals and machinery. Creative tax legislation and other measures that protect farmers' land from development and allow them to pass it on intact to their children. And others, nearly all of which have the virtue of having been proved to work under some circumstances in some places and the liability of leaving many of conventional agriculture's worst problems unchallenged.

In the middle of the spectrum of alternatives to conventional agriculture are a variety of schemes that have at least some environmental component—that is, they take into account the complex relationships among crop plants, farm animals, soils, pests and disease organisms, natural enemies of pests, environmental variables such as climate, and—where appropriate and not too difficult—the cultural heritage of traditional farmers. The best-known of these schemes is integrated pest management, IPM for short, which uses a variety of biological techniques to control insects, together with judicious, carefully timed pesticide applications, as a substitute for heavy aerial and ground spraying of pesticides on an inflexible schedule.

Other mid-spectrum strategies include crop rotation and diversification, the use of "green manure" (plowed-under cover crops of clover or alfalfa to increase soil nitrogen without chemical fertilizer), ridge-tilling for better weed management and control of soil moisture, and the raising of farm animals outdoors rather than in close confinement in "factory" units, to eliminate the need for chemical feed supplements.

All of these strategies have had striking successes in commercial practice. One of the most dramatic has been in the cotton fields, where IPM has allowed U.S. cotton farmers to decrease their pesticide use by more than three-fourths while maintaining yields, lowering costs, and increasing their profit margin. Typical of the newer strategies is one known as "genetically diverse monoculture," which simulates polyculture—the planting of several crops together to confuse insect pests and retard the spread of crop diseases—but uses only one crop. For example, four different genetic varieties of white potato can be planted in the same field, deterring harmful insects but avoiding the harvesting problems of true polyculture. Many more of these mid-spectrum ecological strategies are in the offing. Nevertheless, even the established practices have had their failures: they work well with some crops in some regions, less well in other circumstances, especially in places where conventional farms predominate and pesticide mists drift in the air.

The radical end of the spectrum consists of a few truly revolutionary ecological ideas still in the research stage, the most notable being Wes Jackson's dream of herbaceous perennial polyculture. To understand this dream we have to look back at the kinds of crops and farming methods developed by the first farmers, which set a pattern that Jackson would like to break.

Beginning approximately nine thousand years ago, but at different times in different places (Anatolia, the Levant, part of what is now Iraq and Iran, Africa, and Central and South America), people began to practice agriculture. Most of the food they raised was derived from various species of weedy annual grasses—wheat, corn, rice, millet, sorghum, rye, oats, barley. Some of these crops, such as corn and wheat, were enormously modified from their wild ancestors. These annual grasses still constitute much of our food. Since the invention of agriculture, there has been no change in the basic system of annual plowing and sowing of grain, and no new major crops developed, even during this age of science.

The advantage of annual grains as food is that annual plants, which die after one growing season, invest much of their season's growth in seed production, the only way that they can carry on into the next generation. We eat the seeds. But the disadvantage is that annual seeds must be planted again each year; a great deal of the drudgery and risk of agriculture is based on plowing, planting, weeding, and cultivating young plants—the never-ending cycle of farming.

Wes Jackson would break this cycle by planting a mixture of perennial grasses, perennial legumes, and other perennial plants with edible seeds or fruits. The mixture would imitate the wild prairie, which lives indefinitely without human help. Some of the potential forerunners of crops of this system—eastern gama grass, Illinois bundleflower, Maximilian sunflower—already have been identified.

The tremendous advantages of herbaceous perennial polyculture are evident. The most obvious disadvantage is that it hasn't been invented yet, not in a practical sense. We don't know whether perennial plants can be found and genetically improved so that they will grow together and keep on producing harvestable levels of nutritious seeds year after year without replanting. Many perennial crops are already in the agricultural repertoire: apples and walnuts come to mind. But this is still a far cry from what Wes Jackson first envisioned and has worked hard to achieve; though trees are perennial they do not yield grain.

Farmers and Mountain Sheep— Culture Limits Technology

My worry about what parts of the spectrum of agricultural innovation will replace the Green Revolution has nothing to do with the basic science and engineering of the change. That's another problem. For even if agroecologists can piece together a truly new agriculture from the middle and radical end of the spectrum of innovation, and even if it works, will farmers accept it? Will it spread through the farming community?

Farmers are often conservative and suspicious of new technologies. In fact the behavior of farmers is similar to that of mountain sheep in two critical respects. First, farmers, even corporate farmers, tend to be bound to particular pieces of land for long periods of time. It is a goal of nearly all farmers to be able to transmit their farms in good working condition to their children and grandchildren. Secondly, agri*culture,* the practice of farming, as its name states, is in large part passed like the land itself from generation to generation, from older to younger farmers. As in the case of the mountain sheep, a stable habitat and transmission of information about surviving in that habitat are tightly linked: the cultural transmission of agricultural knowledge is predicated on a continued and long-lasting association with particular pieces of land.

Modern conventional agriculture, the Green Revolution, has ignored this association, substituting material inputs such as chemical fertilizers, farm machinery, pesticides and irrigation for older farming practices such as crop rotation and communal labor systems, and substituting the generalized information of agricultural extension agents, commercial magazines supported by agribusiness, and chemical manufacturer's instructions, for the particular, land-bound cultural information given from parent to child or neighbor to neighbor. Economic and political forces beyond the farmers' control enforced this substitution. Now, with conventional agriculture near collapse, in the nick of time along comes the promise of a new system, collectively called sustainable agriculture. Constructing this new system with the resources and time left to us is a great challenge to our scientific and economic ingenuity—but it is not the only challenge.

My fear is that we will be so carried away with the intellectual thrill of designing a new agricultural technology that we will ignore another challenge: the challenge of making our new system accessible to farmers so that they can use it and further develop it and improve it themselves, get it to provide unique and particular answers to questions about their unique and particular land, and transmit the system to their neighbors and their children. Without this transfer of initiative, sustainable agriculture will be no more successful or durable than the Green Revolution, and it could be considerably worse. If we merely substitute the cult of the benevolent ecologist for the cult of the benevolent pesticide company representative, then

I see little hope for farmers, who have taken about all the benevolence that they can stand.

The new agriculture is one in which informational inputs are being widely substituted for material inputs. Monitoring of pest population sizes to determine the most effective time to take action, and using various techniques to increase enemies of pests can replace much spraying of pesticides. Crop diversification along with imaginative marketing can replace the costly storage facilities needed to store the harvest of single-crop farms. Open-pasture rotation systems for livestock can replace both the antibiotics and food supplements needed when pigs and chickens are raised in factory-style confinement units, which are themselves expensive. New, ecologically based crop-management practices can replace or eliminate herbicides, fungicides, and irrigation pumps.

What appears to be emerging is a new, complex technology, drawing on disciplines as diverse as ecological mathematical modeling, the chemistry of natural products, the modern economic theory of craft production systems, and even genetic engineering. Because the developing technology is complex, new, and highly variable from crop to crop and place to place, I imagine that a whole new farming education industry will arise to service it. Elements of this industry are already in place. Instead of selling chemicals, it sells videotapes, computer hardware and software, reference and networking services, laboratory services, and data analysis services. None of this is necessarily bad, and it could be very exciting; it could also spell the end of a noble experiment and become yet another in the lengthy series of disillusionments and letdowns for modern farmers. This latter possibility must and can be avoided, although to do so will not be easy.

The central scientific defect of the Green Revolution is that it assumes no limits in a world that ecologists know very well to be filled with limiting forces and subject to the unpredictable events that so often bring these limits into play. Knowing this, the one limit that ecologists and other architects of sustainable agriculture cannot afford to ignore is the limit of the ability of human culture to adapt to new technologies, no matter how clever these technologies are or purport to be. The protective conservativism of farmers rightly makes them distrustful of adopting a consumer approach to farming, regardless of how elegant that approach. In other words, they are nervous about depending upon black box techniques of farming whose inner workings they do not understand and whose failures they cannot repair.

Sustainable agriculture has the potential of being presented to farmers the way computer hardware and software are often presented: as a user-friendly package whose technology does not have to be understood. The temptations to present it in this way will be great. After all, as anyone who has tried to understand the mathematics of pest population dynamics or the economics of market systems is likely to agree, sustainable agriculture is not a simple business. But difficult as it may be, the designers of the new agriculture will have to resist the temptation to prepare simplified user-manuals that don't offer serious explanations of what is going on and that don't encourage farmers to modify and improve the system in light of their own practical experience. The experts should welcome the challenge to communicate with farmers in a reciprocal fashion, for once the method of the exchange has been worked out, research and discovery will proceed at a faster rate. And without this kind of communication we are bound to fail, much as modern medicine fails when it treats the patient as a consumer who has no right to participate in the design of his or her own recovery. No technology that treats the farmer as the last and bottom link in a hierarchically organized, expert-dominated chain of transmitted wisdom has a chance of success.

I want to make clear that I do not regard the limitation in the ability of agriculture to adapt to new technology as a defect. It is no more a defect than the tendency of young mountain sheep to follow older ones is a defect. Rather, it is an adaptive response to a complex and risk-laden environment and has great survival value. Sustainable agriculture will have to learn to make use of it, not fight it.

The value of this adaptive response is captured in a statement that was made to me by the Kentucky farmer and writer Wendell Berry: "An adequate relationship between people and the land is not a monologue but a conversation. If the land has something to say back, the good farmer hears it." And this is true even if what the land is saying contradicts the current wisdom of sustainable agriculture. The system has to have room for the farmer's innovative responses to the circumstances and needs of his or her land, a demanding skill that is usually acquired over many years, starting in childhood. Gene Logsdon, an Ohioan with great knowledge about the running of small farms, told an anecdote that illustrates this point in an essay entitled "The importance of traditional farming practices for a sustainable modern agriculture":

> I once asked an Amish farmer who had only twenty-six acres why he didn't acquire a bit more land. He looked around at his ten fine cows, his sons hoeing the corn with him, his spring water running continuously by gravity through house and barn, his few fat hogs, his sturdy buildings, his good wife heaping the table with food, his fine flock of hens, his plot of tobacco and acre of strawberries, his hand-made hickory chairs (which he sold for all the extra cash he really needed), and he said: "Well, I'm just not smart enough to farm any more than this *well.*"

In an essay entitled "Whose head is the farmer using? Whose head is using the farmer?" Wendell Berry made a similar point:

> The good farmer, like an artist, performs within a pattern; he must do one thing while remembering many others. He must be thoughtful of relationships and connections. . . . The good farmer's mind, like any other good mind, is one that can think, but it is by that very token a mind that cannot in any simple way be thought *for*. The good farmer's mind as I understand it, is in a certain critical sense beyond the reach of textbooks and expert advice. Textbooks and expert advice, that is, can be useful to this mind, but only by means of a translation—difficult but possible, which only this mind can make—from the abstract to the particular.

One of the things Berry is saying is that because farming is done in a particular place, the value and application of expert, general advice, even benign ecological advice, is restricted and limited. Green Revolution agriculture tried to get around this dilemma by denying the importance of place, by bypassing place altogether. A factory-farm confinement facility for chickens is much the same whether it is in Iowa or England. Generalized varieties of wheat are expected to grow equally well in Pakistan and Mexico. Agroecologists know better; they reject placeless farming and the troubles it brings. They know that soil types, habitat types, landforms, and microclimates, all varying from place to place, are critical limiting factors.

On the other hand, agroecologists are modern scientists, and at the heart of modern science is—inescapably—the power of generality. Earlier, I described the philosophical problems caused by generality when biologists try to rally round the cause of protecting biological diversity. Agroecologists face an even worse Catch-22. They recognize the central importance of particularity and place in the practice of farming, but their experiments, like any in science, have to control extraneous variables and give nice, general answers.

Reconciling Culture and Technology

How are we to bridge the gap between the applied but general technology of the new agriculture and the daily experience of what Wendell Berry called "the good farmer"? To begin, the experts of sustainable agriculture have to develop new ways of both talking to and listening to farmers. So far, there has been little effort in this direction.

The problem is how to influence farm culture without ruining it—how to create a system in which farmers and experts can interact in a way that does not condescend to the farmers or deny the individuality of their problems, and at the same time maintains the integrity and usefulness of the advice that the experts have to offer. Including farmers as senior partners in the new sustainable agriculture will need as much thought and research as the development of the ecological and economic system itself. Although we still do not know how the farmer and farm culture will be integrated into the new technology of farming, some elements of the eventual solution are beginning to appear in outline, if not in detail.

First, designers of sustainable agriculture—including farmers—have to resist any system that separates farmers from their land. We should learn from the negative example of the futurists, with their seductive fantasy of a farmer sitting at home in front of a computer console while the computer accesses countless data bases and organizes all farm work, from the management of farm machinery to the marketing and accounting, as part of one vast operating program. This fantasy is absurd even for a business or a household—how much more foolish for a farm, where the health of the land and the health of the farmer both depend on the farmer's presence in the fields. During a visit to Shropshire, I had the opportunity of speaking to one of the most respected and successful farmers in the district, not far from Shrewsbury. He complained bitterly that he found it necessary to spend many hours in front of the computer, dealing with the complex economics of the Common Market. For him, the joy of being on his land had been sacrificed to the need to follow the daily fluctuations in the prices of beets in France and farm machinery in Belgium. The way to make farming pleasurable and rewarding is not to "free" the farmer to do something else.

A second element of a proper sustainable agriculture is sure to be the rediscovery of traditional farm wisdom and its incorporation in the new system. Gene Logsdon has called for "an intensive investigation of what traditional farming has learned by trial and error over centuries of experience." Logsdon noted that out of the recognition of this wisdom and the survival value that it confers, Scandinavian countries have subsidized and protected their small, traditional farms and are being amply repaid for this policy by assurance of food sufficiency during times of crisis.

In America, much of the traditional wisdom of farming has been lost or is in danger of dying with the elderly farmers who still remember it. The cause of this tragic loss has been the breakdown of the normal parent-to-child cultural transmission that occurred on many farms during the years following the introduction of the Green Revolution, a period marked by contempt for traditional wisdom. This leads to the third requirement of successful farming: adult education for farmers. Adult education will also be a key element in the relationship between experts and farmers. There is precedent for this sort of education: for example, in a book called *The Small Community*, published in 1942, Arthur Morgan described a rural community in Kentucky where a group of about twenty small farmers met one evening a week to study grape culture.

No one expects that adult education will by itself repair the damaged fabric of farm culture. But as a new culture emerges from the debris left behind after decades of Green Revolution-style farming driven by corporate greed and materialism, the experts must be on hand, locally, to help organize the sharing of useful knowledge. The logical facilitators of this sharing should be the farm extension agents. As Wendell Berry has said, "The good extension agent would be a lexicon of local solutions, not just a conduit from university to farmer.'' The current extension service model of bringing the word from the smart professor to the yokel is outmoded and should be scrapped. Extension agents have to spend more time finding ways to learn and evaluate local information, and helping to spread the best of it horizontally, among farmers.

Fourth, and most difficult to achieve, will be the complete revamping of the philosophy and organization of agricultural research, most of which is long concluded by the time the farmer hears about it. Farmers have to be allowed to help plan the research before it becomes a fait accompli, and in some cases to participate in it as well. This will be difficult because few farmers are qualified scientists and because there is a danger of compromising the creativity and freedom of the researchers. But it is not impossible. In Wisconsin, a group of farmers drew up a research agenda for ecologically oriented field experiments that were subsequently begun at the University of Wisconsin's Platteville campus. On-farm research, in which the private farm becomes a laboratory of the agriculture school or university, is another idea whose time is overdue.

Any truly sustainable agriculture has to be more than a set of technical prescriptions, even ecological ones. Like all successful and durable technologies, it will have to be supportive of the culture that practices it, and participate in that culture. The challenge of changing farming is only one example, albeit the most important one, of the problems we will face as we seek to undo the damage that the outmoded technology of the twentieth century has done to us and to the world. Similar problems will arise as we grapple with housing, medical care, transportation, and law. In each case, the recipients of the new technology will have to be respected, enlightened, and empowered by the experts who serve them. The revolution in agriculture can be an inspiring example of beginning again, provided that everyone remembers three things: Not to separate farming from farmers; not to separate farmers from the particular land on which they farm; and not to imagine that evolution is a process that is directed from above.

Life in the New Millennium

The future is shy. If you want to catch a glimpse of it, you have to sneak up from behind. So the place to start for a look into the future is the past. The question of what life will be like for plants, animals, and people in the new millennium is not a fit subject for casual futurology. We have to lay a good foundation, and we can begin with the relatively recent idea of changing the earth.

Indeed, the idea that we humans can change profoundly the physical and biological world is not old; many people who consider themselves altogether modern still have not grasped it. That the earth can change, most but not all of us accept: we see the work of earthquake, volcano, drought, and flood. Most cultures have a story of a great flood that wiped out a vast number of plants, animals, and human beings.

People, except perhaps those living the most unruffled lives in the most sedentary societies during the most peaceful eras, have been aware for a long time that the world around them can change, often drastically and suddenly. Even religious traditions, which depend for their transmission on unchanging beliefs and values, admit the fact of change. In the Jewish morning prayer service, for example, there are the words "in Your goodness You renew the work of creation every day, constantly." On a less pleasant note are the words in Deuteronomy, chapter 11: "Take heed to yourselves, lest your heart be deceived, and you turn aside and serve other gods, and worship them; and the anger of the Lord be kindled against you, and He shut up the heaven so that there shall be no rain, and the ground shall not yield her fruit; and you perish quickly from off the good land which the Lord gave you." The New Testament also has plenty of discussion of change, especially the familiar verses from the book of Revelation concluding with the words of chapter 21: "Then I saw a new heaven and a new earth; for the first heaven and the first earth had passed away, and the sea was no more."

None of the changes that I have listed and that various people have believed in for a long time—a deluge, the renewal of creation, the shutting of the heavens, the end of the sea, and the passing away of our familiar heaven and earth themselves—are supposed to be the work of human beings, although human action may trigger the change. As I indicated, the idea that people, without divine intervention, can work vast changes in the earth is comparatively new. I do not mean that nobody ever thought of it or thought it possible until now—of course they did—but not until the nineteenth and twentieth centuries did it become a dominant idea of any major society.

Naturally there is no single moment when such a great transition in human thinking takes place. Nevertheless, if I had to pick one day most symbolic of this change I would choose November 17, 1869, the day on which Napoleon's dream of seventy years earlier came true and the Suez Canal was opened to shipping, shortening at one stroke the sea route from England to India by six thousand miles and forever changing the functional geography of the Old World. The dream of Suez, the dream of power and control, of progress, has increasingly dominated our thoughts since that time.

The new idea emerged slowly—slowly enough so that by the time it became commonplace, many of its side effects and problems had already been anticipated. In a way, the problems of the dream of Suez were foreseen in Christopher Marlowe's *Doctor Faustus,* which appeared at the close of the sixteenth century. But the problems were more directly anticipated just four years before the opening of the Suez Canal, in 1865, at the end of the American Civil War, when a book by George P. Marsh was published in New York. Marsh, a diplomat, student of linguistics, and geographer, entitled his work *Man and Nature; or Physical Geography as Modified by Human Action.* On the title page is a quotation from a famous sermon by the Protestant minister Horace Bushnell:

> Not all the winds and storms, and earthquakes, and seas, and seasons of the world, have done so much to revolutionize the earth as MAN, the power of an endless life, has done since the day he came forth upon it, and received dominion over it.

There then follow 549 pages describing major changes that people have worked on the earth and its creatures, and the consequences— good or bad—of these changes. In this first great work of modern conservation, Marsh made it plain that there was no longer any place on earth free of human influence. He wrote about large-scale human operations that "interfere with the spontaneous arrangements of the organic or inorganic world," and he suggested "the possibility and the importance of the restoration of disturbed harmonies and the material improvement of waste and exhausted regions." Man is, Marsh observed, "both in kind and degree, a power of a higher order than any of the other forms of animated life, which like him, are nourished at the table of bounteous nature."

Marsh was expressing an idea that had first been suggested by Charles Babbage nearly thirty years earlier in *The Ninth Bridgewater Treatise,* that anything a human being or animal does, even it it is just to make a sound or to take a step, leaves a permanent, never-dying impression on the processes and physical substrate of the Earth. Marsh took Babbage's idea to a logical next step. Because humans exert so much more power than other animals, the lasting impressions we leave on the planet are much greater than theirs. Consequently, whenever we make some permanent change in the earth, disturb some harmony, it is our responsibility to make some other change that will restore the harmony. Thus the idea of conservation is an evitable outgrowth of the discovery that we can change our planet and that this change can be for the worse.

Looking at Marsh's work, we can see that he was not only the founder of the idea of conservation, at least in America, but also the founder of a particular school of conservation—the interventionist, managerial school, which takes our disruptive presence in the natural world for granted and makes no judgments about it. Later members of the school included Teddy Roosevelt, Gifford Pinchot, and many others.

But meanwhile, another school of conservation thought was forming, setting the stage for a confrontation that continues today. In April of 1868, when Marsh's book had been out three years and while Ferdinand de Lesseps was finishing the Suez Canal, a Scottish-born inventor, wanderer, and naturalist named John Muir was making his first visit to California. Like Marsh, Muir perceived the profound ability of humans to change and damage nature, but his response was different. Here is what Muir wrote about it in his book *The Yosemite,* ten years after his first visit.

> Arriving by the Panama steamer, I stopped one day in San Francisco and then inquired for the nearest way out of town. "But where do you want to go?" asked the man to whom I had applied for this important information. "To any place that is wild," I said. This reply startled him. He seemed to fear that I might be crazy, and therefore the sooner I was out of town the better, so he directed me to the Oakland ferry.

Guiding himself by means of a pocket map, Muir headed east, toward the Yosemite Valley. He continued his account:

> Looking eastward from the summit of the Pacheco Pass one shining morning, a landscape was displayed that after all my wanderings still appears as the most beautiful I have ever beheld. At my feet lay the Great Central Valley of California, level and flowery, like a lake of pure sunshine forty or fifty miles wide, five hundred miles long, one rich furred garden of yellow Compositae. And from the eastern boundary of this vast golden flower-bed rose the mighty Sierra, miles in height, and so gloriously colored and so radiant, it seemed not clothed with light, but wholly composed of it, like the wall of some celestial city.

Muir could not wait to leave San Francisco: the only kind of city in which he could thrive was a city without people, the "celestial city" of the High Sierra. It was Muir more than anyone else who established the other great school of conservation, in which conservation became synonymous with rejection of change and with the preservation of wilderness, through the erection of fences to protect nature from all but the most passive human presence.

One fact remains absolutely clear. Despite their differences, Marsh and Muir were both right in their deeply felt perception that human beings can have and increasingly were having a profound and lasting effect on the natural world. More than a century later, few people yet comprehend the magnitude of that effect. Large numbers alone without a recognizable context have little meaning. According to the Worldwatch Institute, thanks to destructive agricultural practices the earth is losing twenty-three billion tons of topsoil annually, even taking into account new soil formation. But what does twenty-three billion tons mean? Considering the size of the earth, is it a lot or a little?

Here is a more comprehensible statistic: on April 3, 1924, at the Agricultural Research and Education Center in Belle Glade, Florida, a nine-foot, graduated concrete post was sunk straight down through the rich muck soil of this drained part of the Everglades until it hit bedrock and the top of the post was at ground level. The area was fenced off. By April 1979, fifty-five years later, the forces of oxidation working on soil that had been drained for an inappropriate kind of agriculture had lowered the ground level so that five feet of that post were exposed. In this protected spot, less than half the soil remained. This is the soil that in nearby parts of South Florida once produced most of our winter vegetables. In some places in Florida, that soil cover today measures only a few inches over the limestone bedrock and is becoming impossible to farm. Many winter fruits and vegetables eaten in the United States now come from Mexico and even Australia, and will continue to do so until their soils also give out or transportation becomes too expensive—which ever comes first.

In the former Soviet Union, the Aral Sea in central Asia, a huge body of water with an average depth of fifty feet in 1965, is shrinking because the rivers that once fed it with fresh water have been entirely diverted for agricultural irrigation. Unless the diverted water is somehow replaced, by the year 2000 the Aral Sea, once the world's fourth-largest lake, will no longer exist—it will have been replaced by salt marsh and salt desert.

In El Salvador in 1961, there were approximately a half-million acres of forest and woodland. By 1978, the total forested acreage of El Salvador was close to zero. It had all been turned into pasture for beef cattle for fast-food hamburgers in the United States.

Naturally the worldwide extinction of species parallels the scale of these sorts of incredible habitat destructions and alterations. Once, exasperated by some people who were claiming that extinction was nothing to worry about because it was only a natural process, not a sign of human-caused change, I compared the modern extinction rate among most groups of mammals with the extinction rate that occurred during the last great die-off, in the ice ages of the Pleistocene epoch. The extinction rate is now, conservatively calculated, at least a thousand times greater. Recent calculations by various ecologists are more sophisticated than mine but not more encouraging. One estimate has it that by the time the world's primary tropical forests are two-thirds cut—in the first decades of the new millennium—as many as 625,000 species will have become extinct. Other estimates are considerably higher. Only 1.7 million species of plants and animals have been named to date.

We are ending the twentieth century confronted by the loss of many of the species in whose company we began the century. In group after major group of plants and animals we find a significant percentage, sometimes a majority, occasionally all of the species endangered. Among the less-studied groups, we have undoubtedly lost species that we never got a chance to name or even to recognize. For the first time, we are seeing the terrifying, sudden disappearance of the familiar building blocks of nature. All over the world frogs are vanishing, inexplicably. In Europe, the only continent where the fungi are well known, we have finally noticed that the mushrooms are going, and with them the forests whose tree roots cannot absorb nutrients from the soil alone. In the face of these monumental changes caused by humans—numberless and often arcane changes—neither fencing off land nor actively managing it will slow significantly the rate of decline. Neither Muir nor Marsh gives us the remedy.

Fences and gamekeepers to keep people out of protected areas may have worked in the days of King Henry VIII; they do not work now. I think of the Hutcheson Forest of Rutgers University, a sixty-acre piece of pristine, precolonial forest in central New Jersey. Never cut, never plowed. The last major fire in 1711. Used only for unobtrusive ecological research. The public kept out except for closely guided tours on a single narrow trail, once every two to four weeks. And what is the forest like? The canopy has great gaps crisscrossed by trailing vines of grape, poison ivy, and bittersweet, a sure sign of severe disturbance. Nonnative species are everywhere: Norway maple, *Alianthus* (the tree of heaven), Japanese honeysuckle, the egg cases of gypsy moths. You can fence out people, but you cannot fence out their effects. And this is a place where there are no rhinoceroses and elephants to attract the horn and ivory poachers, no exotic, tropical orchids to tempt unscrupulous plant collectors; it should be easy to protect. But alien introduced pests, acid rain, ozone, insecticide residues, drifting herbicides, heavy metals, atmospheric particulates—these effects and creations of our society can be anywhere and everywhere on earth.

Not only the effects of people but also the people themselves are more pervasive and invasive than previously. There are more people than there ever were before. Those with leisure and money have scubas and snowmobiles; even the poor have chain saws. Where can people be kept from overrunning nature? Remote islands and top-security military bases are fairly well protected, yet they are only a small percentage of the earth's surface.

There is another problem with protection. In ecosystem after ecosystem, ecological studies are beginning to show that a protected oasis in a sea of development loses species quickly. How big does a park or preserve have to be to save its species? This question is only now being asked in different parts of the world, and the answer is always the same: as big as you can get, but it won't be big enough.

If protection is a weak reed, active management and intervention are not much better, despite our technological self-confidence. Consider as an example of this type of conservation the preservation of animal species in zoos or special breeding facilities and the preservation of agricultural plant varieties in seed banks. William Conway, the director of the New York Zoological Society, and George Rabb, director of the Chicago Zoological Park, both effective conservationists, computed in 1980 that the care of the 750 Siberian tigers in the world's zoos would cost $49 million just to maintain them for twenty years, until the year 2000. Five hundred gorillas would cost $47 million. If at that time we had saved just two thousand selected species out of the tens and probably hundreds of thousands that are endangered, it would cost $25 billion by the year 2000, said Conway, who, I should point out, was advocating the effort. Moreover, this is only the economics of it—never mind the genetic worries about whether species will become irreversibly changed in zoos or the ecological problem of the disappearance of their native habitat while they are in captivity.

Seed banks face similar difficulties—they are expensive, ecologically and evolutionarily nonsensical, and vulnerable to the loss of entire collections through accidents such as power failures, through sabotage, or through bureaucratic misunderstandings. Seed banks are inadequate to preserve more than a small fraction of even the recorded varieties of crop plants. At Kew Gardens in England, it was estimated that new acquisitions were just balancing the seeds being lost from storage. The situation at the United States' major seed bank, at Fort Collins, Colorado, appears to be even worse. This is not an encouraging record.

Beyond zoos and seed banks, lies the ultimate dream of the management-oriented conservation of species: genetic engineering. We will create new species to replace the old ones, some say. But recreating from scratch just one species would be a feat that would make the stuffing of animals into Noah's ark simple by comparison. Barring a miracle, genetic engineering is simply not worthy of serious consideration as a conservation tool.

If species cannot be reconstructed, ecosystems can. "Restoration ecology" is a vigorous new field, with solid, exciting successes to its credit. Wastelands and dumps have been turned back into prairies and salt marshes—even some tropical dry forests are being reconstructed from disconnected fragments of the old habitat. But this task is enormous in scope, each piece of land poses unique challenges, the restoration is never perfect (often not even close), and restored small patches in a larger devastated landscape are of limited value and staying power.

Before going on, I want to stress that we must continue and redouble all our legitimate efforts in both protection and management to resist the destructive changes we are working on nature. We need the restoration projects, the parks, preserves and wilderness areas, the seed banks, the zoos, and the aquariums in spite of their weaknesses. They will not be adequate to save the world's endangered species and habitats if things continue on their present course, but suppose, as is likely, that our course changes. What then? True, the state of affairs might get worse, but it also might get better, and in that event any interim conservation that we will have achieved will be of value.

At this point, I want to say that I believe that *the ultimate success of all our efforts to stop ruining nature will depend on a revision of the way we use the world in our everyday living when we are not thinking about conservation.* If we have to conserve the earth in spite of ourselves, we will not be able to do it.

Here is an example—a modern parable of constructive living. In the Papago Indian country of Arizona's and Mexico's Sonoran Desert, beautifully described in Gary Nabhan's book *The Desert Smells Like Rain,* there are two similar oases only thirty miles apart. The northern one, A'al Waipia, is in the U.S. Organ Pipe Cactus National Monument, and is fully protected as a bird sanctuary, with no human activity except bird-watching allowed. All Papago farming, which was practiced there continuously since prehistoric times, was stopped in 1957. The other oasis, Ki:towak, over the border in Mexico, is still being farmed in traditional Papago style by a group of Indian villagers.

Nabhan, an ethnobotanist, visited the oases "on back-to-back days three times during one year," accompanied by ornithologists. He found fewer than thirty-two species of birds at the Park Service's bird sanctuary but more than sixty-five species at the farmed oasis. Asked about this, the village elder at Ki:towak replied:

> I've been thinking over what you say about not so many birds living over there anymore. That's because those birds, they come where the people are. When the people live and work in a place, and plant their seeds and water their trees, the birds go live with them. They like those places, there's plenty to eat and that's when we are friends to them.

And that is when the destructive changing of nature ceases, when people who are not actively trying to save the world play and work in a way that is compatible with the existence of the other native species of the region. When that happens—and it happens more than we may think—the presence of people and the changes they bring may enhance the species richness of the area, rather than exert the negative effect that is more familiar to us.

I have earlier noted that the only way to predict impending change and its consequences is to project trends and processes that are already clearly present. This approach to prediction is not as adventurous as science fiction or the more glamorous kinds of futurology; however, I hope that it is more solid and realistic. Having said this, I can now consider some of the possible futures we may face in the days ahead.

First and most obvious is the possibility—not a likely one, I think—that during the next twenty-five to fifty years things will go on pretty much as they have since the 1950s. In other words, there will be more people, there will be more industrialization, there will be more urban growth, there will be more standardization, there will be more corporate conglomeration and bigger organizations, there will be more power-oriented consumer goods, there will be more tourism, there will be huge military budgets, there will be more advertising and image making, there will be less and less room for personal eccentricities, there will be more mechanization of agriculture and chemical farming, there will be more managers.

If this is the course of events during the next half-century, then it becomes a fairly easy job to predict the fate of species and habitats on earth. In my textbook, *Biological Conservation,* I had a section that I titled "Characteristics of Endangered Species." In it was a table with two columns, one labeled "Endangered" and the other labeled "Safe." In the table I listed the sorts of characteristics that might put a species at high or low risk of extinction, and I gave contrasting examples using related species of animals. Here is a sample of the characteristics I listed in the table. Endangered species of animals are likely to be of large size, such as the mountain lion; safe species are likely to be small, such as the wildcat. Species that have a restricted distribution—an island, a few bogs, a desert spring—are often endangered; the Puerto Rico parrot is an example. Species with wide distribution, such as the yellow-headed Amazon parrots, are at least comparatively better off. Species that are intolerant of the presence of humans, the grizzly bear for example, are endangered; tolerant species, such as the black bear, are safe. Species with behavioral idiosyncracies that are not adaptive in urban or suburban areas do poorly: the redheaded woodpecker flies from one tree to the next in a low, swooping arc that often brings it into the path of oncoming cars. On the other hand, some species have, by chance, behavioral patterns that suit them for coexistence with us. The burrowing owl is tolerant of noise and has a kind of flight that lets it evade oncoming objects. Burrowing owls have been able to survive between the runways of the Miami International Airport.

Having listed the characteristics of endangered species of animals, I went on to construct a hypothetical "most endangered animal." It is a large predator with a narrow habitat tolerance, long gestation period, and few young per litter. Although hunted for a natural product or for sport, it is not subject to proper game management. It has a restricted distribution but travels across international boundaries. It is intolerant of humans or their pollution, reproduces in large, vulnerable groups and has behaviors that are poorly adapted for survival in populated regions. There may be no such animal, but with a few qualifications the description might fit the Mediterranean monk seal.

Conversely, if you take the opposite characteristics—small size, herbivorous diet, high fecundity, wide distribution, and so forth— you get a composite picture of the typical wild animal of the twenty-first century. Some of the most familiar existing approximations would be the house sparrow, the gray squirrel, the Virginia opossum, and the Norway rat.

We could also include in that list the common pigeon or rock dove, the familiar species of domestic cockroach, the feral house cat, and such creatures as the rapidly spreading eastern (U.S.) canid, which is part dog and part coyote with maybe a little bit of wolf mixed in. Among the plants, in the twenty-first century we will see more of *Ailanthus,* which thrives in areas covered with asphalt and concrete; ragweed, which does best in recently disturbed soils; and the tall reed grass *Phragmites,* which loves wet spots, regardless of how much air and water pollution is in the neighborhood.

These are the jolly companions we can expect if the world continues much longer on its present course. The tropical forests will be gone, along with their myriad animal species; the temperate forests will be going. There will still be rhinoceroses in zoos, showy orchids in botanical gardens, and, in seed banks, slowly or rapidly deteriorating poorly representative collections of what is left of the human agricultural heritage: a few varieties of African upland rice, more varieties of wheat, a smattering of beans. The fruit trees will fare even worse, although I suppose the awful Red Delicious apple will be around forever as a perpetual reminder of our sins.

That is the future as I see it, if we simply extend what is happening now. No conservation efforts, either protective or managerial, will reverse this basic trend toward extinction, although we will have a few victories here and there, now and then. I do not mean to imply that the only species that will be left will be the weeds, the pests, and the vermin, although these are certainly the species that do best in the damaged ecosystems that will prevail. There will still be salutary species but far fewer of them than there are now. We are selecting for the tough species, the resilient species, the species of upheaval, and these tend to be the ones that we do not like and that are not good for us. As it says in the superficially simpleminded statement in Deuteronomy, chapter 30, "Choose life that you may live." Conversely, if we continue to choose a culture that survives through destruction and death, we should not be surprised if destruction and death are what we get.

As for human society in this scenario, simply read the newspapers and extrapolate. It is safe to assume that cities will not get smaller, crime and drug abuse will not lessen, pollution will not abate, and life will not be more fulfilling for most people everywhere if we are unable or unwilling to leave the path we are now on.

There is one part of modern life that many people feel will completely and favorably alter the course of events without necessitating any change in our beliefs or objectives. I refer to the so-called information revolution. This revolution, made possible by computers and their software, has occurred in three areas: first, the storing, retrieving, and manipulating of information; second, communications; and third, the redesigning of the structure and function of organizations, which is just beginning. To what extent will these developments modify the prognosis for species, habitats, and society in the world of the future? Will our mastery of information enable us to sidestep the problems caused by the rest of our social organization and technology?

The late Jesuit scholar Pierre Teilhard de Chardin foresaw the day when all human consciousness would flow together into one great unified layer, becoming part of the ''noosphere'' and enveloping the earth in its collective intellectual and spiritual wisdom. Have we moved, are we moving, in that direction? I think not.

Those who predict major, even unlimited benefits from the information revolution ignore the existence of certain fundamental limits inherent in it: the limit to the value and usefulness of pure information is one; the limit to the quality of information available is another; the limit to how much we can manipulate human culture and social institutions to meet the needs of computers is the last. These limits interact in many ways to restrict severely the possibilities of the information revolution, and we are already starting to see that. I will give only a few illustrations of the problems without trying to identify the particular limits that caused them.

What I am talking about goes beyond the GIGO (garbage in, garbage out) principle. We are foundering in a sea of information. One of the most profound and funniest science fiction stories of Stanislaw Lem is found in his book *The Cyberiad*. The heroes, Trurl and Klapaucius, the ''constructors'' I mentioned in a previous chapter, are trapped inside their spaceship in a remote junkyard corner of space by a pirate named Pugg, a hideous robotic monster with a Ph.D. Pugg wants information even more than gold, so Trurl and Klapaucius design and construct for him a variant of the perpetual motion machine, a gadget that generates random facts about the universe at incredible speed and then selects and prints out those that are true.

The tiny diamond-tipped pen shivered and twitched like one insane, and it seemed to Pugg that any minute now he would learn the most fabulous, unheard-of-things, things that would open up to him the Ultimate Mystery of Being, so he greedily read everything that flew out from under the diamond nib . . . the sizes of bedroom slippers available on the continent of Cob, with pompons and without . . . and the average width of the fontanel in indigenous step-infants . . . and the inaugural catcalls of the Duke of Zilch, and six ways to cook cream of wheat . . . and the names of all the citizens of Foofaraw Junction beginning with the letter M, and the results of a poll of opinions on the taste of beer mixed with mushroom syrup. . . .

And it grew dark before his hundred eyes, and he cried out in a mighty voice that he'd had enough, but Information had so swathed and swaddled him in its three hundred thousand tangled paper miles, that he couldn't move and had to read on about how Kipling would have written the beginning to his second *Jungle Book* if he had had indigestion just then, and what thoughts come to unmarried whales getting on in years and all about the courtship of the carrion fly . . . and why we don't capitalize paris in plaster of paris.

And many more things that I won't include. Meanwhile, Trurl and Klapaucius escape, leaving Pugg to his hideous punishment, which is also recorded somewhere on the endless stream of tape coming out of the information machine.

One of the most troublesome aspects of the information revolution is its tendency to promote overmanagement—I call this the Real Goods Problem. We cannot eat information, we cannot wear it, and information will not take out the garbage at night. As an increasing percentage of citizens becomes involved with the manipulation of information, much of it worthless, fewer people are left to concern themselves with the real goods and services needed for carrying on life. It can be argued that the information manipulators are increasing the efficiency of everyone else. But again there is a limit to what can be accomplished by increasing the efficiency of designs and processes, and efficiency may have unexpected and undesirable effects. These effects include disrupting the lives of people and communities, and—less obvious—making it easier for us to do the wrong thing. Our century has been slow to grasp the truth that the efficiency of an action has absolutely nothing to do with its rightness or wrongness.

What is happening in the information revolution is a massive and complicated trade-off of information for personal and social stability—a trade-off of clever assumptions for reality. The trade-off is sometimes beneficial and sometimes harmful, but it is certainly not leading us toward the paradise promised by the revolution's promoters. So I do not think that the information revolution is likely to modify the sad forecast for the fate of nature and human society if things continue on as they have in the recent past. As we speed along on the express track of modern technological inventiveness, headed toward the promised brilliant future, we hear a roar and feel the pull as another train rushes by in the opposite direction. Look quickly, and we will see ourselves seated there, too, on the express called the Reality of Life, headed not into the sunrise but into the storm and the gathering darkness.

If we continue on that track, sooner or later we will find that it ends in global chaos—upheaval and breakdown on a vast scale. Before the fragmentation of the Soviet empire, in 1991, nuclear war was the prime threat of global chaos, one that few discussed and nearly everyone feared. But the Soviet empire is in shreds and the American empire, nominally intact, has been taken over by its creditors as the United States sinks into a bottomless pit of debt. Huge nuclear arsenals are too expensive to keep, and unless there is a slip—always possible—most of the nuclear bombs will be dismantled gingerly and somehow disposed of, possibly by burning some of the radioactive contents in nuclear power plants. A few bombs will be kept and others will be developed by smaller nations, including nations controlled by amoral dictators and religious fanatics. Thus the danger of "limited" nuclear war remains, albeit lessened somewhat. If it happens, and if it goes much beyond the exploding of two or three weapons, the changes that will occur in nature and human society do not bear thinking about, at least not by me.

A more likely cause of upheaval is the disintegration of the extremely complicated and finicky economic, industrial, social, and political structure that we have put together in the decades since the Second World War. This structure has been supported by resources, especially petroleum, that are waning, and by an environmental and cultural legacy—soil, vegetation, air, water, families, traditions—that we foolishly took for granted, squandered, and lost.

The visible agent of this change would be a global economic collapse. Such a collapse would probably disrupt international trade, trigger the disintegration of many multinational corporations and other overstuffed, subsidized superorganizations, end the modern welfare state, diminish governmental regulatory supervision (including environmental regulation), bring about massive famines and movements of populations, greatly increase unemployment in the industrialized nations, all but eliminate luxury goods and exceedingly complex manufactures, including many advanced military weapons, hasten the spread of new and old epidemic diseases, trigger the inevitable population crash, and cause a proliferation of regional economic, social, and political systems. Some of these things are already happening.

In America and similar societies, these changes will be particularly dramatic. I can only mention a few of the more noteworthy ones. There will be major unemployment or reduced employment not only of blue-collar workers but of managers, professionals, and people who work in the "service sector" (travel and real estate agents, operators of fitness centers, securities salepersons, and so forth). The employment problem will be aggravated by the contraction of the largest high-technology enterprises, especially those subsidized directly or indirectly by the government. These include NASA, military research, large universities, much of agribusiness, and the medical research establishment. Many of the people who will be affected, especially the managers, have no useful skills and will suffer greatly.

The Age of Global Control is coming to its inevitable end. It is like a massive flywheel, spinning too fast for its size and construction, coming apart in chunks as it spins. There goes a chunk—the sick and aged along with the huge apparatus of doctors, social workers, hospitals, nursing homes, drug companies, and manufacturers of sophisticated medical equipment, which service their clients at enormous cost but don't help them very much.

There go the college students along with the vice presidents, provosts, deans, and professors who have not prepared them for life in a changing world after formal schooling is over.

There go the high school and elementary school students, along with the parents, school administrators, and frustrated teachers who have turned the majority of schools into costly, stagnant, and violent babysitting services.

There go the lawyers and their hapless clients, in a dust cloud of the ten billion codes, rules, and regulations that were produced to organize and control an increasingly intricate, unorganizable, and uncontrollable society.

There go the economists with their worthless, pretentious predictions and systems, along with the unemployed, the impoverished, and the displaced who reaped the consequences of theories and schemes with faulty premises and indecent objectives.

There go the engineers, designers and technologists, along with the people stuck with the deadly buildings, roads, power plants, dams, and machinery that are the experts' monuments.

There go the advertising hucksters with their consumer goods, and there go the consumers, consumed by their consumption.

And there go the media pundits and pollsters, along with all those unfortunates who wasted precious time listening to them explain why the flywheel could never come apart, or tell how to patch it even while increasing its crazy rate of spin.

It is not hard to imagine some of the practical consequences of the breakdown, and I see little point in dwelling on them. But a few images stick in my mind. I see huge, unsellable, suburban houses with attached three-car garages, abandoned by their owners of record, scavenged for usable parts and contents, surrounded by wild lawns filled with dandelions, brambles, dying ornamental plants, and vigorous, deep-rooted, ungainly tree of heaven saplings. I see endless shelves and display cases of "collectibles," which—like exercise machines—were a way of burning up excess wealth, and which will have reassumed their rightful monetary value: nothing.

The most terrifying thing about this disintegration for a society that believes in prediction and control will be the randomness of its violent consequences. The chaotic violence will include not only desperate, ruthless struggles over the wealth that remains, but the last great rape of nature. What will make it worse is that, at least at the beginning, it will take place under a cloud of denial and cynical reassurances. We will be told that "War Is Peace, Freedom Is Slavery, Ignorance Is Strength," even after Big Brother is dead and his empire a shambles. The denial and reassurance have already begun—I don't know how many people are being fooled. Yet even some of those who aren't, who are not inextricably bound up in the sytem, who see the flywheel breaking up and are preparing to take cover, will be struck down by random chunks: knowledge of what is happening will help but will not be a guarantee of safety in the midst of chaos.

In the undeveloped world, many of these processes of decline are well underway with brutal effect. In a growing number of countries public services such as transportation and medical care have all but stopped, and the cities, swollen by millions of refugees displaced from the rural areas by the technology and hidden agendas of export agriculture, are experiencing unprecedented populations, crime, and misery. That this will get much worse as the industrial world collapses is evident, at least for the short run. Eventually, if nations and regions are left to work their problems out separately, in isolation, things will get better for some—there is no predicting which ones.

And what of nature in the undeveloped world? I believe that a global economic collapse will have two different and opposite effects on world floras and faunas. In countries such as India, Bangladesh, Mexico, and some of the more densely populated of the African nations, where natural resources, especially trees, are already seriously depleted, little of the native flora and fauna will survive. As export industries and cash crops fail, unprepared and disrupted local agricultural communities will have to feed themselves and large urban populations—an impossible burden on human resources and on eroded and poisoned soils. Anyone who flies over Mexico today will see what an obvious prediction this is. On the other hand, international economic collapse would also bring about the cessation of many disruptive processes. No longer would Japanese industry and companies like Weyerhauser and Georgia-Pacific be cutting the last remaining primary forests of Australia and the South Pacific nations at today's staggering rate. No longer would Volkswagen and Liquigas and the like be using defoliants or napalm or bulldozers to turn Amazonian rain forest into short-lived cattle ranches for the fast-food trade. No longer would there be a market for pet hyacinthine macaws and palm cockatoos, at five thousand dollars or more apiece. So although global economic breakdown would strain the ecosystems and species of some places past the breaking point, it would give ecosystems and species in other places a new lease on life. And in many areas these contradictory processes would be happening side by side.

We cannot begin again as if the twentieth century had not happened, as if it had caused no lasting damage to the physical world, to nature, or to human culture. Nor would we want to reject those true advances in the human condition that have come from our social and technical inventiveness. Any utopian vision that pretends that nothing irreversible happened in the years before the new millennium is a lie. Nevertheless, there is a second alternative to mitigate the unrelieved bleakness of the first. Even in the midst of upheaval and disintegration it is possible to change course. Eventually, perhaps quickly, a different way of living will be organized, a better system will form. Some of its possible outlines are coalescing and becoming visible even now. This second alternative already exists within and as a response to the Age of Control, it is growing, and it may yet come to a blessed dominance.

This second alternative is a transformation of the dream of progress from one of overweening hubris, love of quantity and consumption, waste, and the idiot's goal of perpetual growth to one of honesty, resilience, appreciation of beauty and scale, and stability—based in part on the inventive imitation of nature. We have already had examples of what this alternative can be like: the *chinampas,* or swamp gardens, that were the glory of pre-Columbian Mexican farming and which might again sustain the Mexican people; the city of Florence in the Renaissance and the city of Toronto before the building boom of the 1970s and 1980s; the hedgerows of post-Elizabethan England; old Jerusalem and the terraces of the Judean hills; and the ingenious multicrop gardens of tropical west Africa, to name a few. The changes that people inevitably work on the earth do not have to be destructive ones.

If this alternative way of living grows and prospers, I doubt that it will do so by some master plan or protocol. Instead, it will be advanced by countless people working separately and in small groups, sharing only a common dream of life. They will tend to be flexible, inventive, and pragmatic, and most will have practical skills—carpentry, the building of windmills and small bridges, the design and repair of engines and computers, the recognition and care of soils, the ability to teach. Nature will have entered their lives at an early age and will remain as a source of joy and as the measure of their best and worst efforts. They will welcome the challenge of the world that Orwell hoped for, a simpler, harder world in which machines, like their inventors, are understood to be limited. They will devote their first energies to the places where they live. They will come to authority not by violence but by their evident ability to replace a crumbling system with something better. And they will share an awe for a power nobler and larger than themselves, be it God, nature, or human history.

And then they must remember what happened before their time. The Jewish observance of the Passover, the exodus from Egypt, can serve as a model of the kind of remembrance that brings survival. At the heart of the Passover is a family ceremonial meal, the seder, a Hebrew word which means order. The seder, which may last five or six hours or more, is built around an orderly and largely prescribed retelling of the story of the exodus in song and narrative. Everyone who is old enough to speak participates. There are no absolute heroes or villains in the retelling: Moses is mentioned only twice; the death of the Egyptian host in the sea is observed with sadness, not glee, by the deliberate spilling of wine from a full cup. The original participants in that ancient drama are spoken of as "I" or "we," not "they"; everyone at the seder is meant to feel that he or she was there. Only in this way can the significance of great historical events be recreated with accuracy and relevance for each successive generation.

In similar fashion, the twentieth century must be remembered by those who follow, even by that majority that will want to forget. The leaders who eventually guide society back to a true heading will be the most fit to tell the story and establish the ritual of its remembrance. They should not dwell on heroes and villains but on the pleasures and sufferings of the age that was left behind, on the causes of its deadly impermanence, and on the choice—life—that inevitably determined what was kept and what was discarded by the founders of the new millennium.

Suggested Readings

The following list includes both sources cited in the preceding pages and some useful readings that were not mentioned. Many items are relevant to more than one chapter, but I have listed each item only once.

Places

Wendell Berry. *A Place on Earth*. San Francisco: North Point Press, 1983.

Archie Carr. *Ulendo: Travels of a Naturalist In and Out of Africa*. New York: Knopf, 1964.

_____. *The Windward Road*. Tallahassee, Florida: University Presses of Florida, 1979 (1955).

Stan Rowe. *Home Place: Essays on Ecology*. Edmonton, Alberta: NeWest (No. 204, 8631–109 St., Edmonton, Alberta. T6G IE8, Canada), 1990.

Scott Russell Sanders. *Home Ground*. Boston: Beacon Press, 1993.

Peter Sauer, ed. *Finding Home: Writing on Nature and Culture from* Orion *Magazine*. Boston: Beacon Press, 1992.

Henry David Thoreau. *Cape Cod*. New York: Viking Penguin, 1987 (1855).

The Roots of Prophecy:
Orwell and Nature

Jaquetta Hawkes. *A Land*. Boston: Beacon Press, 1991 (1951).

George Orwell. *Down and Out in Paris and London*. New York: Harcourt Brace Jovanovich, 1972 (1933).

_____. *A Clergyman's Daughter*. New York: Harcourt Brace Jovanovich, 1969 (1935).

_____. *The Road to Wigan Pier*. New York: Harcourt Brace Jovanovich, 1972 (1937).

_____. *Coming Up for Air*. New York: Harcourt Brace Jovanovich, 1969 (1939).

_____. *1984*. New York: Signet/New American Library, 1983 (1949).

_____. *The Collected Essays, Journalism and Letters of George Orwell*, 4 vols., Sonia Orwell and Ian Angus, eds. New York: Harcourt Brace Jovanovich, 1968.

Michael Sheldon. *Orwell: The Authorized Biography*. New York: Harper-Collins, 1991.

The Rain Forest of Selborne

William Bartram. *The Travels of William Bartram: Naturalist's Edition*. Francis Harper, ed. New Haven: Yale University Press, 1958 (1791).

William Beebe. *Jungle Peace*. With a Foreword by Theodore Roosevelt. New York: Henry Holt, 1926 (1918).

Charles Waterton. *Wanderings in South America*. London: Thomas Nelson & Sons, 1900 (1825).

Gilbert White. *The Natural History and Antiquities of Selborne, In the County of Southampton: With Engravings and an Appendix*. London: Benjamin White & Son, 1788–89. (The first edition.)

_____. *The Natural History and Antiquities of Selborne, In the County of Southampton*. Thomas Bell, ed., 2 vols. London: John Van Voorst, 1877. (A nineteenth-century edition edited and annotated by a biologist who lived in White's former house in Selborne.)

_____. *The Natural History of Selborne*. Richard Mabey, ed. Harmondsworth, Middlesex: Penguin Books, 1977. (A good twentieth-century edition.)

Desert Life

Edward Abbey. *Desert Solitaire: A Season in the Wilderness*. New York: Simon & Schuster, 1968.

Wendell Berry. "Getting along with nature," in *Home Economics*. San Francisco: North Point Press, 1987, pp. 6–20.

Charles M. Doughty. *Travels in Arabia Deserta,* 2 vols., 3rd ed. New York: Dover, 1979 (1921).

Michael Evenari, Leslie Shanan, and Naphtali Tadmor. *The Negev: The Challenge of a Desert.* Cambridge, Massachusetts: Harvard University Press, 1971.

Daniel Hillel. *Negev: Land, Water, and Life in a Desert Environment.* New York: Praeger, 1982.

Max Oelschlager. *The Idea of Wilderness.* New Haven: Yale University Press, 1991.

Sigurd Olson. *Reflections from the North Country.* New York: Knopf, 1976.

A Turtle Named Mack

Dr. Seuss. *And To Think That I Saw It On Mulberry Street.* New York: Random House, 1989 (1937).

——————. *The 500 Hats of Bartholomew Cubbins.* New York: Random House, 1989 (1938).

——————. *The King's Stilts.* New York: Random House, 1939.

——————. *Bartholomew and the Oobleck.* New York: Random House, 1949.

——————. *Yertle the Turtle and Other Stories.* New York: Random House, 1950.

——————. *The Lorax.* New York: Random House, 1971.

The Overmanaged Society

Loren Baritz. *Backfire: A History of How American Culture Led Us Into Vietnam and Made Us Fight the Way We Did.* New York: Ballantine Books, 1985.

Charles Dickens. *Little Dorrit.* New York: Penguin Books, 1967 (1857).

John Fowles. *The French Lieutenant's Woman.* New York: Signet/New American Library, 1970.

Primo Levi. *Survival in Auschwitz.* New York: Collier Books/ Macmillan, 1961 (1959).

Lewis Mumford. *The City in History: Its Origins, Its Transformations, and Its Prospects.* New York: Harcourt, Brace & World, 1961.

——————. *The Pentagon of Power (The Myth of the Machine,* vol. 2). New York: Harcourt Brace Jovanovich, 1970.

C. Northcote Parkinson. *Parkinson's Law*. Boston: Houghton Mifflin, 1957.

Paul Von Ward. *Dismantling the Pyramid*. Washington, D.C.: Delphi Press, 1981.

Forgetting / State of the Art

Erwin Chargaff. *Heraclitean Fire: Sketches from a Life before Nature*. New York: The Rockefeller University Press, 1978.

John Dewey. *The School and Society*. Carbondale, Illinois: Southern Illinois University Press, 1980 (1899).

Carolyn Jabs. *The Heirloom Gardener*. San Francisco: Sierra Club Books, 1984.

Bill McKibben. *The Age of Missing Information*. New York: Random House, 1992.

David W. Orr. *Ecological Literacy: Education and the Transition to a Postmodern World*. Albany, New York: State University of New York Press, 1992.

Page Smith. *Killing the Spirit: Higher Education in America*. New York: Viking Penguin, 1990.

Alfred North Whitehead. *The Aims of Education and Other Essays*. New York: Free Press/Macmillan, 1967 (1929).

Bruce Wilshire. *The Moral Collapse of the University*. Albany, New York: State University of New York Press, 1990.

The Lesson of the Tower

Hannah Arendt. *The Human Condition*. Chicago: University of Chicago Press, 1958.

David Lavery. *Late for the Sky: The Mentality of the Space Age*. Carbondale, Illinois: Southern Illinois University Press, 1992.

Walter A. McDougall. . . . *The Heavens and the Earth: A Political History of the Space Age*. New York: Basic Books, 1985.

Arnold Pacey. *The Culture of Technology*. Cambridge, Massachusetts: The MIT Press, 1983.

After Valdez

Charles Perrow. *Normal Accidents: Living with High-Risk Technologies*. New York: Basic Books, 1984.

Eugene S. Schwartz. *Overskill: The Decline of Technology in Modern Civilization.* New York: Ballantine Books, 1971.

Dent de Lion—The Lion's Tooth

Jeff Cox. *Landscaping with Nature: Using Nature's Designs to Plan Your Yard.* Emmaus, Pennsylvania: Rodale Press, 1991.

Alastair Hay. *The Chemical Scythe: Lessons of 2,4,5-T and Dioxin.* New York: Plenum Press, 1982.

Geoffrey Jellicoe, Susan Jellicoe, Patrick Goode, and Michael Lancaster. *The Oxford Companion to Gardens.* New York: Oxford University Press, 1986.

K. Schneider. "Senate panel says lawn chemicals harm many," in *The New York Times,* May 10, 1991.

Edwin Rollin Spencer. *All About Weeds.* New York: Dover, 1974 (1957).

William K. Stevens. "A lawn fed on chemicals switches to an organic diet," in *The New York Times,* August 27, 1991.

_____. "With a little cheating, organic lawn turns 2," in *The New York Times,* September 1, 1992.

The Technology of Destruction

William J. Broad. *Teller's War: The Top-Secret Story Behind the Star Wars Deception.* New York: Simon & Schuster, 1992.

The Bulletin of the Atomic Scientists. Published by the Educational Foundation for Nuclear Science, Inc., 6042 S. Kimbark Ave., Chicago, Illinois 60637. (An excellent source of balanced information on the military, space, and nuclear power technologies. For the general reader.)

The Defense Monitor. Published by the Center for Defense Information, 1500 Massachusetts Ave. NW, Washington, D.C. 20005. (An authoritative newsletter about weapons and military programs compiled largely by retired military officers.)

H. Freedman and M. Simon, eds. and trans. *Midrash Rabbah,* Vol. VIII: *Ruth, Ecclesiastes.* London: Soncino, 1939.

The Koran. N. J. Dawood, trans. Harmondsworth, Middlesex: Penguin Books, 1974, Sura 30: "The Greeks (Al Rum)." (The quotation cited includes wording from several translations.)

Norman Lamm. "Ecology in Jewish Law and Theology," in *Faith and Doubt.* New York: Ktav, 1971, pp. 162–185.

Robert Jungk. *Brighter Than a Thousand Suns: A Personal History of the Atomic Scientists.* New York: Harcourt, Brace & World, 1958.

Stockholm International Peace Research Institute (by Arthur H. Westing). *Ecological Consequences of the Second Indochina War.* Stockholm: SIPRI/Almqvist & Wiksell International, 1976.

——————. *Weapons of Mass Destruction and the Environment.* London: Taylor & Francis, 1977.

——————. *Warfare in a Fragile World: Military Impact on the Human Environment.* London: Taylor & Francis, 1980.

Hard Times for Diversity

David Ehrenfeld. *The Arrogance of Humanism.* New York: Oxford University Press, 1981.

——————. "The management of diversity: A conservation paradox," in *Ecology, Economics, Ethics: The Broken Circle,* F. Herbert Bormann and Stephen R. Kellert, eds. New Haven: Yale University Press, 1991, pp. 26–39.

Paul R. Ehrlich and Anne H. Ehrlich. *Healing the Planet: Strategies for Resolving the Environmental Crisis.* Reading, Massachusetts: Addison Wesley, 1991.

Charles S. Elton. *The Ecology of Invasions by Animals and Plants.* London: Methuen, 1958.

Aldo Leopold. *The River of the Mother of God and Other Essays by Aldo Leopold,* Susan L. Flader and J. Baird Callicott, eds. Madison, Wisconsin: University of Wisconsin Press, 1991.

Bryan Norton. *Why Preserve Natural Variety?* Princeton, New Jersey: Princeton University Press, 1987.

Rights

Samuel Butler. *Erewhon and Erewhon Revisited.* New York: Modern Library, 1927 (1872, 1901).

Aldo Leopold. *A Sand County Almanac: With Other Essays on Conservation from Round River.* New York: Oxford University Press, 1966.

Christopher D. Stone. *Should Trees Have Standing? Toward Legal Rights for Natural Objects,* 2nd ed. Portola Valley, California: Tioga, 1988.

Loyalty

Inner Voice. The newsletter of the Association of Forest Service Employees for Environmental Ethics, P.O. Box 11615, Eugene, Oregon 97440.

Gifford Pinchot. *A Primer of Forestry: Part II—Practical Forestry* (Bulletin No. 24, Part II, USDA Bureau of Forestry). Washington, D.C.: U.S. Government Printing Office, 1905.

George W. S. Trow. "The Harvard Black Rock Forest," in *The New Yorker*, June 11, 1984, pp. 44–99.

Trustees of Dartmouth College v. Woodward; 4 Wheaton, 518. In *Documents of American History*, 5th ed., Henry Steele Commager, ed. New York: Appleton-Century-Crofts, 1949, pp. 220–223.

Ecosystem Health

John J. Berger. *Environmental Restoration*. Washington, D.C.: Island Press, 1990.

Robert Costanza, Bryan Norton, and Ben Haskell, eds. *Ecosystem Health: New Goals for Environmental Management*. Washington, D.C.: Island Press, 1992.

Steward T. A. Pickett and Peter S. White, eds. *The Ecology of Natural Disturbance and Patch Dynamics*. San Diego, California: Academic Press, 1985.

David W. Schindler. "Detecting ecosystem responses to anthropogenic stress," in *Canadian Journal of Fisheries and Aquatic Science*, Vol. 44, 1987, pp. 6–25.

Down from the Pedestal—
A New Role for Experts

Herman E. Daly and John B. Cobb, Jr. *For the Common Good: Redirecting the Economy Toward Community, the Environment, and a Sustainable Future*. Boston: Beacon Press, 1989.

P. A. Larkin. "An epitaph for the concept of maximum sustained yield," in *Transactions of the American Fisheries Society*, Vol. 106, 1977, pp. 1–11.

Mary Midgley. "Why smartness is not enough," in *Rethinking the Curriculum: Toward an Integrated, Interdisciplinary College Education*, Mary

E. Clark and Sandra A. Wawrytko, eds. Westport, Connecticut: Greenwood Press, 1990.

David W. Orr. See his "Conservation Education" columns in the journal *Conservation Biology,* beginning with the March, 1989 issue (Vol. 3, No. 1, p. 10). The quotations in this chapter come from "Politics, conservation, and public education," Vol. 5, No. 1, March 1991, pp. 10–12.

E. F. Schumacher. *Small Is Beautiful: Economics As If People Mattered.* New York: Harper & Row, 1989 (1973).

Asking Unaskable Questions—A Case Study

Carrett Hardin. "The tragedy of the commons," in *Science,* Vol. 162, 1968, pp. 1243–1248.

——————. *Stalking the Wild Taboo,* 2nd ed. Los Altos, California: William Kaufmann, Inc., 1978.

——————. *Naked Emperors: Essays of a Taboo Stalker.* Los Altos, California: William Kaufmann, Inc., 1982.

Changing the Way We Farm

Wendell Berry. *The Unsettling of America: Culture and Agriculture.* San Francisco: Sierra Club Books, 1977.

——————. *The Gift of Good Land.* San Francisco: North Point Press, 1981.

——————. "Whose head is the farmer using? Whose head is using the farmer?" in *Meeting the Expectations of the Land,* Wes Jackson, Wendell Berry, and Bruce Coleman, eds. San Francisco: North Point Press, 1984, pp. 19–30.

Masanobu Fukuoka. *The One-Straw Revolution.* Emmaus, Pennsylvania: Rodale Press, 1978.

Valerius Geist. *Mountain Sheep: A Study in Behavior and Evolution.* Chicago: University of Chicago Press, 1971.

Wes Jackson. *New Roots for Agriculture,* 2nd ed. Lincoln, Nebraska: University of Nebraska Press, 1985.

——————. *Altars of Unhewn Stone.* San Francisco: North Point Press, 1987.

David Kline. *Great Possessions: An Amish Farmer's Journal.* San Francisco: North Point Press, 1990.

Gene Logsdon. "The importance of traditional farming practices for a sustainable modern agriculture," in *Meeting the Expectations of the Land, op. cit.,* pp. 3–18.

Arthur Morgan. *The Small Community.* New York: Harper, 1942.

Judith D. Soule and Jon K. Piper. *Farming in Nature's Image: An Ecological Approach to Agriculture.* Washington, D.C.: Island Press, 1992.

Angus Wright. *The Death of Ramón González: The Modern Agricultural Dilemma.* Austin, Texas: University of Texas Press, 1990.

Life in the New Millennium

Charles Babbage. *The Ninth Bridgewater Treatise,* 2nd ed. London: Frank Cass, 1967 (1838).

William Conway. "Zoos, Future Directions," in *Animal Kingdom,* April-May 1980, pp. 28–32.

David Ehrenfeld. *Biological Conservation.* New York: Holt, Rinehart, and Winston, 1970. See also: David Ehrenfeld. *Conserving Life on Earth.* New York: Oxford University Press, 1972.

Stephen Fox. *The American Conservation Movement: John Muir and His Legacy.* Madison, Wisconsin: University of Wisconsin Press, 1985.

Al Gore. *Earth in the Balance: Ecology and the Human Spirit.* Boston: Houghton Mifflin, 1992.

Jane Jacobs. *Cities and the Wealth of Nations: Principles of Economic Life.* New York: Random House, 1984.

Les Kaufman and Kenneth Mallory, eds. *The Last Extinction.* Cambridge, Massachusetts: The MIT Press, 1986.

Christopher Lasch. *The True and Only Heaven: Progress and Its Critics.* New York: W. W. Norton, 1991.

Stanislaw Lem. *The Cyberiad: Fables for the Cybernetic Age,* Michael Kandel, trans. New York: Avon, 1985.

Barry Lopez. *The Rediscovery of North America.* New York: Vintage, 1992.

George P. Marsh. *Man and Nature; or, Physical Geography As Modified By Human Action.* New York: Charles Scribner, 1865.

John Muir. *The Yosemite.* Madison, Wisconsin: University of Wisconsin Press, 1987 (1912).

Gary Paul Nabhan. *The Desert Smells Like Rain.* San Francisco: North Point Press, 1982.

Orion: People and Nature. A magazine published by The Myrin Institute, Inc., 136 E. 64th St., New York, New York 10021. (Contains the most thoughtful environmental writing now being published.)

Theodore Roszak. *The Cult of Information: The Folklore of Computers and the True Art of Thinking.* New York: Pantheon, 1986.

State of the World. Published annually by the Worldwatch Institute (and W. W. Norton), 1776 Massachusetts Ave. NW, Washington, D.C. 20036. See also Worldwatch Papers, monographs on selected environmental topics. (The most comprehensive source of information on the environmental crisis.)

Pierre Teilhard de Chardin. "The antiquity and world expansion of human culture," in *Man's Role In Changing the Face of the Earth,* William L. Thomas, Jr., *et al.,* eds. Chicago: University of Chicago Press, 1956, pp. 103–112.

Index

A'al Waipia, 183
Abortion, 158
Acid rain, 139, 156, 181
Ad Dibdibdah, 34
Administration. *See* Management
Administrators. *See* Bureaucrats; Management
Advanced Publications, Inc., 134
Advertising, 184, 191
Africa, 166, 186, 192–93
Agent Orange, 102
Agnew, Harold, 56
Agribusiness, 168, 190
Agriculture, 26–27, 38, 70, 150, 163–74, 179, 184, 192; culture of, 169–74, 186
Agroecologists, 167, 171–74
AIDS (acquired immune deficiency syndrome), 129
Ailanthus (tree of heaven), 181, 185, 191
Air traffic control, 148, 155
Alaska, 95–96
Alfalfa, 166
Algae, 91, 149
Alien species. *See* Exotic (nonnative) species
Allah, 112
Allee, W. Clyde, 140, 143
Alligator, 3–4, 133
Amazonian rain forest, 192

American Fisheries Society, 148
Amish, 171
Anatolia, 166
Animal rights. *See* Rights
An Nafūd, 34
Annual crops, 166–67
Anthropocentrism, 108
Antibiotics, 53, 115, 169
Anti-Semitism, 10
Apollo moon program, 61, 83, 85, 87–88
Apple, 59–60, 167; Cox's Orange Pippin, 17, 24; Red Delicious, 186
Aquariums, 183
Aral Sea, 179
Arcania australis, 128–30
Archdiocese of New York, 57
Arizona, 38, 183
Armstrong, Neil, 90
Ar Rab' al Khālī, 34
Asia Minor, 101
Asimov, Isaac, 90
Association of Forest Service Employees for Environmental Ethics (AFSEEE), 137
Assumptions, false, x–xi
Astronauts, 86–87
Atomic Energy Commission, 106
Auschwitz, 58
Australia, 130, 179, 192
Aviation industry, 82, 87

Babbage, Charles, 177
Babel, 79–80, 82, 94
Balance of nature. *See* Equilibrium, natural
Baldwin, Ernest, 67–68
Bangladesh, 161, 192
Barley, 166
Barnard College, 74
Barrington, Daines, 30–32
Bateson, Gregory, 117
Bauer, Raymond, 93
Beans, 186
Bears; black, 185; grizzly, 149, 185
Bedouins, 38
Beebe, William, xi, 32
Bees; Euglossine, 107; honey, 133
Beetles, 121
Beets, 173
Belgium, 173
Bella, David, 55–56
Berger, John J., 139
Berkeley. *See* University of California
Berry, Wendell, xii, 112, 170–73
Bethe, Hans, 106
Bible, xi, 79, 82, 116, 122
Big Brother, 191
Biochemistry, comparative, 67–68
Biosphere, 110, 121, 123
Bison. *See* Buffaloes
Bittersweet, 181
Blair, Eileen, 15, 20
Blair, Eric. *See* Orwell, George
Blair, Richard, 20, 22
Blake Hall, Old, 73–74
Blake, Maurice A., 73
Blake, William, 47, 125
Board of Education, New York City, 57
Bohr, Niels, 104
Bomb, atomic. *See* H-bomb; Weapons, nuclear
Bormann, F. Herbert, 156
Botanical gardens, 123, 129
Botticelli, 114
Bougainvillea, 21–22
Bratton, Susan, 131
Britain, 22, 25, 59–60
British Coal Board, 154

Brogdale Horticultural Station, 59–60
Brownell, Sonia, 22
Buddhism, 92–93; Buddhist economics, 154
Buffaloes, 128, 130–31
Bulldozers, 192
Bundleflower, Illinois, 167
Bureaucracy. *See* Bureaucrats; Management
Bureaucrats, 28, 52, 57, 62, 95, 99, 127, 156, 182, 184, 190. *See also* Management
Burnham, James, 8
Bush, George H., 85
Bushnell, Horace, 177
Butler, Samuel, 128

Cactuses, 36
California, 103, 105–6, 121, 155, 158, 178
Camel, 36–37
Canadian government, 130
Canid, eastern (U.S.), 185
Capital, 63
Capitalism, 25–26
Carp, 13
Carr, Archie, xii, 3–7
Carson, Rachel, 154
Carter administration, 62–63
Cat, feral house, 185
Catastrophe theory, 142
Catch-22, 117, 172
Cathedral of St. John the Divine, 65
Catholic bishops, 112
Central America, 166
Centralization, 11
Cerro de la Muerte, 7
Chaco Canyon, 109
Challenger, 56, 80, 84, 88–89, 93
Chamomile, Roman, 103
Chaos, 64, 160, 189, 191
Chaparral, 121
Chargaff, Erwin, 54, 122
Chelonia mydas. See Turtle, green
Chemicals, farm, 164–66, 168–69, 184. *See also* Fertilizer; Fungicides; Herbicides; Insecticides; Pesticides

Chemistry, 115, 117, 120, 169
Chernobyl, 80, 93, 96
Chickens, 169, 171
Children, 42, 44, 46, 93, 100, 102–3, 119, 122, 162, 168, 173
China, 9
Chinampas, 193
Chinese, 159
Christianity, 92
Chubb, John E., 57
Churchman, C. West, 61, 87
Chutzpah, 93. See also Hubris
Circumlocution Office, 52
Civilization, 133
Claremont College, 155
Clark, Colin, 118
Clements, Frederick E., 140
Climate, change of, ix, 143
Clover, 166
Cobb, John B., Jr., 155
Cockatoos, palm, 192
Cockneys, 19
Cockroach, 185
Collapse, global economic, 189–92
Collectibles, 191
Colonialism, 24–25
Colorado, 182
Columbia University, 63
Columbus, Christopher, 83, 91
Common, Jack, 21
Common Market (European), 173
Commons, 161–62
Communism, 26, 82, 84, 86; anti-, 106
Community; human, 137, 142, 165, 167, 173, 188; natural, 109, 120, 122, 139–44, 164
Computers, 172–73, 186–87, 193
Condor, California, 121
Congress, U.S., 62, 85, 127, 136
Conservation, 3, 6, 33, 95–98, 117, 119, 122, 131, 136, 153, 177–78, 181–83, 186; economics of, 118–21
Conservation biologists, 95, 153
Conservation Biology, xii, 153
Conservative, 63, 159, 169
Consumerism (and consumption), 11, 98, 169, 184, 191, 193

Control (over nature), xi, 27, 52, 64, 91, 100, 103, 105–6, 115, 117, 190–92. *See also* Power (over nature)
Conway, William, 181–82
Copenhagen, 127
Coral reefs, 108
Corn (maize), 166
Corporations; greed of, 173; loyalty to, 132–33, 134, 136; multinational, 189
Coryanthes, 107
Costa Rica, 3–4, 6
Cotton, 166
Coyotes, 149, 185
Creation, 113
Creosote bush, 35
Cretaceous, 5
Crick, Francis H., 123
Crime, 192
Crops, 166–67; diversification of, 166, 169; rotation of, 166, 168; varieties of, 165, 167, 182
Cubbins, Bartholomew, 44–46
Culture; adaptability of, 169–74, 187; heritage of, ix, 165, 168, 173, 189, 192
Cumberland Island National Seashore, 131

Daedalus, 80
Da Goma, Juan, 128–29
Daisies, 103
Daly, Herman E., 155
Dandelions, 99–103, 191
Darwin, Charles, 70
David (King), xi
Death Valley, 35
DeBonis, Jeff, 137
Debt, 189
Deception, 14
Deer, 131
Delaware-Raritan Canal, 41–42, 46
De Lesseps, Ferdinand, 178
Democracy, U.S., 83
Demoralization, 58
Denial, 191
Dentistry, 148, 150, 155
Depression, Great, 142

Desert, 35–39, 141, 179; Arabian, 34–35; life in, 38; Negev, 38–39; Sahara, 35–37; Sonoran, 38, 183; Southwestern, 35
Determinism, technological, 93
Deuteronomy, vii, 113, 175–76, 186
Dickens, Charles, 49, 52
Dirac, Paul, 104
Disasters, 95–98
Disturbance, environmental, 140–44, 181
Diversity, biological, 107–8, 114–23, 142, 148, 150, 153, 172. *See also* Generality; Uniformity; Value (of biological diversity)
Doctor Faustus, 176
Dogs, 133, 185
Domesticated animals, 133
Donkeys, 131
Double bind, 117
Doughty, Charles, 38
Dove, rock, 185
Dr. Dolittle, 90
Dr. Seuss, 42–46
Dubos, René, 109
DuPont, 56

Eagles, bald, 149
Earthquake, forecasts of, 152, 159–60
Earthworms, 71
East Bengal, 161
Ecology (and ecologists), 18, 66, 111, 119–21, 123, 130, 136, 139–46, 148, 150–52, 155–56, 160, 181–82; agricultural, 163, 166–74
Economics (and economists), 117–21, 123, 130, 135, 148, 150, 154–55, 182, 190; agricultural, 168–72; global economic collapse, 189–92
Ecosystems, 33, 95, 111, 119, 182, 186, 192; health of, 139–46; stability of, 142
Ecotourism, 120
Education, 57, 69, 88, 156, 169, 173–74. *See also* Teaching; Universities
Efficiency, 59, 188
Egypt, 39, 194
Ehrlich, Paul, 111

Einstein, Albert, 105
Eisenhower, Dwight, 83–85
Elijah, ix–x, xii, 39
Elisha, ix
Elk, 131
El Salvador, 179–80
Endangered species, 120
Energy, effects of, 109–10
Engineers, 76, 191
England, 13, 29, 176, 193
Enterprise, 90
Eocene, 5
Epidemic diseases, 190
Equilibrium, natural, 142–44
Ethics, 137
Euphorbias, 36
Europe, 82, 101, 103, 180
Evenari, Michael, 39
Everglades, 179
Evolution, 91, 163, 174, 182
Exodus, 9, 194
Exotic (nonnative) species, 130–31, 181
Experts, 11, 68, 76, 98, 147–57, 170–74, 191
Exploitation, 24, 27, 49, 68, 108, 117, 119, 122–23
Extension agents, agricultural, 168, 173–74
Extinction, 111–12, 118, 121, 180, 184, 186
Exxon Valdez, 95–98

Famines, 189
Farming, xi, 39, 101, 120, 163–74. *See also* Agriculture
Fascism, 25–26
Feasibility study, 150
Feedback, positive, 53, 55–56, 58, 62, 70
Fermi, Enrico, 104–5
Fertilizer, 164, 166, 168
Feynman, Richard, 56
Figueres, José, 6–7
Fish, "cleaner," 108
Flathead Lake, 149
Flood, great, 175
Florence, 193

Florida, 3, 179
Flower, 101–3, 107
Flycatcher, 30
Ford, Henry, 132
Ford Motor Company, 132
Forecast, forecasting, 150–52, 154, 159, 188. *See also* Predictions; Prophecy
Forest; Black Rock, 135–37; Harvard, 135–36; Hutcheson, 180; loss of, ix, 120, 122, 139, 180, 186; products, 120; subject to disturbance, 141. *See also* Forest Service (U.S.); National forests (U.S.); Rain forest
Forest Service (U.S.), 132, 136–37
Fort Collins, Colorado (seed bank), 60, 182
Fowles, John, 59
France, 173
Franck, James, 104
Fresh water, loss of, ix
Frogs, disappearance of, 180
Fungi, 119; disappearance of, 180
Fungicides, 164, 169
Fusion, cold, 61
Futurology, 8, 175, 184

Gainesville, Florida, 3
Game ranches, 131
Gamow, George, 104
Ganges delta, 162
Garden; multicrop, 193; of Eden, 113
Geisel, Theodor S. *See* Dr. Seuss
Geist, Valerius, 163–64
Gene bank, vegetable, at Wellesbourne, 60
Generality, 107, 115–16, 123, 172. *See also* Diversity, biological; Uniformity
Genes, 116, 119, 131, 133, 150
Genesis, 82
Genetic engineers, engineering, 59–60, 68, 115–16, 122, 136, 153, 169, 182
Gentleman, Victorian, 59
Geological Survey, U.S., 62
Georgia-Pacific, 192
Germany, German, 104–5
Glacier National Park, 149
Goats, 129–30, 164

God, 9, 11, 23, 39, 91–93, 113, 121–22, 130, 176, 194; imitation of, x
Goddard, Robert, 81
Golden, William T., 135
Gorer, Geoffrey, 21
Gorillas, 181
Grain, 166–67
Grants, research, 55–56, 62, 70–72, 156
Grape, 173; vines, 181
Grass, 99–100, 102–3; annual, 166–67; eastern gama, 167; perennial, 167
Grasslands, 141
Green manure, 166
Green Revolution, 164–65, 167–69, 171, 173
Growth, 53, 61–62, 69, 193; limits to, 128
Gulag, 81
Gypsies, 19
Gypsy moths, 141, 181

Hagelstein, Peter, 110
Haldane, J. B. S., 122
Hampshire, 31, 33
Hanford, Washington (nuclear facility), 56
Hardin, Garrett, 158–62
Harvard University, 3, 135–37
H-bomb, 105–6, 110, 189. *See also* Weapons, nuclear
Health; human, 144–45, 172; of ecosystems, 139–46, 153, 172
Hebrew, 194
Hedge, 17, 193
Heinlein, Robert, 90
Heisenberg, Werner, 104
Helmer, Nora and Torvald, 127
Hemmer, Helmut, 133
Herbicides, 102, 164, 169, 181
Hertz Foundation, 109
Hilbert, David, 104
Hirsch, Samson Raphael, 82
Hitler, Adolf, 104–5
Holocaust, 89, 112
Honesty, 25, 193. *See also* Nature, honesty in
Honeysuckle, Japanese, 181

Hoover Institution (of Stanford University), 106
Hops, 19–20
Horoscope, 152, 154
Horses, 131
Housing, 174
Hubris, xi, 151, 193. *See also Chutzpah*
Hudson River, 135
Hurricanes, 159
Hyde, Rod, 110

Ibexes, 164
IBM (International Business Machines), 132
ICBM (intercontinental ballistic missile), 81, 83
Illusion, 14
Image, 85–87, 89, 93, 136, 184
India, 93, 176, 192
Indian, American, 109, 114; Apache, 103; Digger, 103; Papago, 183
Indian Ocean, 128
Individualistic theory, 142
Information, 127; revolution, 186–88; transmission of, 56, 164, 168–70, 173–74, 187
Inner Voice, 137
Insecticides, 164, 181
Integrated pest management (IPM), 165–66
Internal Revenue Service, 62
International Harvester, 134
International Wildlife Fund, 129
Invertebrates, 121, 149
Iowa, battleship, 65
IPM. *See* Integrated Pest Management
Iran, 166
Iraq, 34, 143, 166
Irrigation, 164–65, 168–69
Isaiah, 28
Israel, xi, 38, 143, 165

Jackson, Wes, xii, 166–67
Japanese, 57–58, 88, 118, 192
Jaques, Eleanor, 21
Jargon, 147
Jefferson, Thomas, 114, 138

Jerusalem, 193
Job, book of, 122
Johnson, Lyndon, 81, 83, 85–86
Josephus, Flavius, 116
Jubilee year, 138
Judaism (Jewish), 92, 104, 112, 116, 175, 194
Judean hills, 193
Jungk, Robert, 105

Kangaroo rats, 36
Kangaroos, 130
Kansas, 141
Kennedy, John F., 81, 83, 85–87
Kentucky, 170, 173
Kerala, 93
Kew Gardens, 182. *See also* Seed banks
Khrushchev, Nikita, 81
King Ahab, ix
King Derwin, 44–46
King Lear, x
Kipling, Rudyard, 188
Ki:towak, 183
Klapaucius, 187–88
Knopf, Alfred A., Inc., 134
Knowledge, loss of, 65–72, 77, 173
Kondratyuk, Yuri, 81
Koran, 112
Korolev, Sergei, 81

Labor, 63
Lake Alice, 4
Lake Erie, 139
Lake trout, 149
Lamm, Norman, 113
Land, needs of, 170–74
Landmarks, natural, x–xi, 44
Larkin, Philip A., 148–49
Law (and lawyers), 174, 190
Lawn, 99–100, 102–3
Lawrence Livermore National Laboratory, 105–6, 109
LeGuin, Ursula, 90
Legumes, perennial, 167
Lem, Stanislaw, 90, 187
Lenin, 82
Leontieff, Wassily, 155

Leopold, Aldo, 128
Levant, 166
Levi, Primo, 58
Leviticus, 138
Lewis, C. S., 8, 11, 90
Liberal, 63, 159, 161
Lichens, 36
Limnologist, 67
Linnaeus, 123
Liquigas, 192
Liverpool, 129
Loach, 123
Locality, 136
Lofting, Hugh, 90
Logsdon, Gene, 170–71, 173
London, 19–20
Los Alamos, 105
Love Canal, 88
Loyalty, 132–38
Lymphoma, 102

Macaws, hyacinthine, 192
Machine, 25–27, 193
Machinery, farm, 164–65, 168, 172–73
Madison, James, 138
Mammal Liberation Front, 129
Management, 49–64, 79, 188; conserva-
    tion, 96, 177, 180, 182, 186; of agri-
    culture, 59–60; of business, 54, 56–
    57, 134, of government, 50–51, 56,
    62; of hospitals, 51–52, 54; of
    schools, 57; of science, 58–61, 156; of
    universities, 53–56, 62–63, 68–72,
    76–77, 134; revolution against, 63–64
Managerial Revolution, 69
Managers. *See* Bureaucrats; Management
Manhattan Project, 105–6
Maple, Norway, 181
March of Dimes, 134
Market, world, 120
Marlowe, Christopher, 176
Marsh, George P., 176–80
Marshall, John, 132–33, 137
Marx brothers, 89
Marxism, 50
McCarthy era, 106
McClelland, B. R., 149

McDougall, Walter A., 80–93
McIlhenny, E. A., 4
McNamara, Robert, 87
Medical care, 174, 192
Mercury program, 86
Mexico, 38, 179, 183, 192–93
Miami International Airport, 185
Middle East, 34
Midrash, xi
Migrant farmworkers, 19
Military, 82, 84–89, 184, 190
Military-industrial complex, 142
Millennium, new, xi, 175, 180,
    193–94
Millet, 166
Ministry of Agriculture, Fisheries and
    Food (MAFF), 59–60
Minks, 149
MIRV (multiple independently targeted
    reentry vehicle), 84
Missouri, 152, 158
Model; or example, 194; predictive, 149–
    50, 152, 155, 169
Moe, Terry, 57
Monoculture, 166, 169
Monod, Theodore, 36
Moose, 163–64
Morgan, Arthur, 173
Moses, xi, 9, 39, 194
Mountain lion, 184
Muir, John, 103, 127, 178–80
Mumford, Lewis, 8, 52, 64, 136
Museums, natural history, 123
Mushrooms. *See* Fungi
Mussels, 98
Mycorrhizal, 119

Nabateans, 38–39
Nabhan, Gary, 183
Napalm, 192
Napoleon, 176
NASA (National Aeronautics and Space
    Administration), 56, 61, 84–85, 87–
    89, 190
National forests (U.S.), 132, 136;
    Willamette, 137
National Park Service (U.S.), 131

National Science Foundation, 62
National Seed Storage Laboratory, 60, 182
Native Americans. *See* Indian, American
Natural history, 32, (72). *See also The Natural History of Selborne*
Natural resources, 49, 137, 157, 189, 192. *See also* Nonresources
Natural Resources Building, 74–78
Nature; and God, 23–24; and people, 18, 26, 31, 33, 175–86, 189, 191–94; as standard and guidepost, 11–12, 28; beauty and serenity of, 11, 18–20, 193; honesty of, 11–14; imitation of, x–xi, 193; love of, 18; reliability/continuity/durability/resilience of, 11, 15–17. *See also* Control (over nature); Power (over nature)
Navistar, 134
Nazis, 25, 104
Newhouse, Si, 134
New Jersey, 3, 5, 33, 57, 101, 180
New Mexico, 109
Niemela, Aina, xii
Nimrod, 79, 82
Nixon, Richard M., 85–86
Nonresources, 120–21
Noosphere, 187
Novak, Michael, 159
Nuclear physics. *See* Physics
Nuclear power (and reactor), 11, 27, 61, 106, 144, 189
Nuclear winter hypothesis, 111
Nutrient loading, 149–50

Oases, 183
Oats, 166
Oberlin College, 153
Obligations, 130–31
Odysseus, 128
O-Group, 110
Ohio, 170
Oil spill, 95–98
Oligocene, 5
Oman, 34
Oobleck, 45–46
OPEC (Organization of Petroleum Exporting Countries), 154

Opossum (Virginia), 185
Oppenheimer, J. Robert, 104–6
Orchids, 107, 186
Organismic theory, 140, 142–43
Oregon, 137
Organizations, 184, 186, 189; loyalty to, 133–35, 137–38
Organ Pipe Cactus National Monument, 183
Orion magazine, xii
Orr, David, xii, 153
Orwell, George, 8–28, 52–53, 99, 193
Ostriches, 128
Otters; river, 149; sea, 96–97
Overhead, 55, 60, 62, 70–71, 156
Overmanagement. *See* Management
Overpopulation, 63, 161–62, 181, 192
Owl, burrowing, 185
Ozone, 181

Pacey, Arnold, 92–93
Paperwork, 50–51, 53–54, 58
Parkinson, C. Northcote, 53
Parrots, 133; Puerto Rico, 184; Yellow-headed Amazon, 184
Partisan Review, 10
Passover, 194
Patents, 55, 117
Pennant, Thomas, 30, 123
Perennial, 101, 103; crops, 167
Pesticides, 166, 168–69
Pests, insect, 102, 165–66, 169–70, 186
Pharmacopoeia, U.S., 103
Phragmites, 185
Physics, 104–7, 111–12, 115
Pickett, Steward, 141
Pigeon, common, 185
Pigs, 131, 169, 171
Pinchot, Gifford, 132, 136–37, 178
Pine Barrens (New Jersey), 101
Pirsig, Robert, 92
Places, awareness of, 3–4, 6–7, 43–44, 46, 123, 136, 171–74
Planners, Soviet, 26, 77
Pleistocene, 180

Plowing, 167
Poachers, 6
Poison ivy, 181
Pollsters, 191
Pollution, 185–86
Polyculture, 166–67, 169
Population; crash, 190; growth, ix, 192; movements, 189. *See also* Overpopulation
Porpoise, 96
Potato, white, 166
Poverty, 19–20, 84, 137, 190
Powell, Anthony, 20
Power (over nature), xi, 52, 59–60, 64, 92, 110, 115–17, 151, 157, 177. *See also* Control (over nature)
Prairie, 141, 167
Predictions, 119, 147–52, 154–55, 159–60, 169, 184, 190–92. *See also* Prophecy
Preserve, nature, 181–82
Prince William Sound, 95
Privacy, 54, 58, 77
Producers (production), 50, 52, 55, 57–58, 62–64, 188
Progress, 25–27, 65, 81, 193
Prophecy, ix–x, 8–11, 27–28
Prophets. *See* Prophecy
Protocooperation, 143
Psychology, educational, 148, 151
Public relations, 84, 86, 88
Pueblo Bonito, 109
Pugg, 187–88
Putnam Rolling Ladder Company, 74

Rabb, George, 181
Ragweed, 185
Rain forest, 33, 120, 192. *See also* Forest
Random House, 134
Rat, Norway, 185
RCA, 134
Reagan, Ronald, 63, 85, 110
Reductionist, 148
Redwoods, 121
Rees, Richard, 15

Reorganization Project (Carter administration's), 62
Research, 55, 59–62, 67–72, 105–6, 141, 153–54, 156, 159–60, 190; agricultural, 170–74
Resources. *See* Natural resources
Restoration, 98; ecology, 139, 182
Revelation, book of, 176
Revolution, 63; scientific-technical (S-T), 81–83
Rhinoceroses, 186
Rice, 166; African upland, 186
Rights, 127–31, 142
Rockefeller Foundation, 104
Rocket (rocketry), 80–81, 86, 88, 91
Roosevelt, Franklin, 105
Roosevelt, Theodore, 32, 178
Rootlessness, 136. *See also* Places, awareness of
Rosewarne Station, 60
Russia, 9, 25, 88, 163
Rutgers University, 73–76, 180
Rye, 166

Salmon; kokanee, 149–50; Pacific, 149–50
San Andreas Fault, 121, (159)
San Francisco, 178
San Joaquin Valley, 103
San José, Costa Rica, 7
Satellites, space, 84, 91
Savannah River nuclear weapons plant, 56–57
Schindler, David, 143
Schmidt-Nielsen, Bodil and Knut, 36–37
Schneider, Keith, 56
Schumacher, E. F., 8, 11, 92, 150, 154–55
Science (scientists), 55, 58–60, 66, 71, 85, 96, 116–17, 122, 145–46, 148, 150, 153–55, 161; agricultural, 167, 171–72, 174
Science fiction, 90, 184
Seals; harbor, 96, 98; Mediterranean monk, 185

Seder, 194
Seed, 101–2, 167
Seed banks, 181–83, 186
Selborne, 29–33, 123
Shakespeare, 114
Shanan, Leslie, 39
Sheep, 130; mountain, 163–64, 167–68, 170
Shepherds, xi
Shkolnik, Amiram, 38
Shrewsbury, 173
Shrimp, opossum, 149
Shropshire, 173
Sierra (mountains), 178
Sin, original, 92
Skills, practical, 193
Sloth, xi
Socialism, 24–27
Soils, 70–71, 101, 120, 166, 171, 180, 193; loss of, ix, 130, 165, 179, 189
Solar design, 109
Sorghum, 166
South America, 166
Soviet Union, 8, 77, 80–81, 83–84, 86, 88, 112, 142, 163, 179, 189
Soyuz, 84
Space, conquest of (and program), 79–93
Sparrow; house, 185; Wrightson giant sea, 128–30
Specialism, 160
Species, endangered, (118–19), 120, (121–23), 182–85
Speciesism, 142
Specificity, 115–17, 123. *See also* Diversity, biological; Generality; Uniformity
Speedwells, 103
Spencer, C. N., 149
Spiders, 36
Sputnik, 80, 84
Squirrel, gray, 185
Stability, 188
Stanford, J. A., 149
Stanford University, 106
Star Wars, 84, 88, 91, 109–10
State Department, U.S., 50–51
Steiner, Daniel, 136

Stevens, William K., 141
Stillman, Ernest G., 135–36
Stillman, John S., 135
Stone, Christopher, 128
Subsidies, crop, 165
Succession (community), 140
Suez Canal, 176, 178
Sunflower, Maximilian, 167
Superconductivity, 61
Supreme Court, U.S., 127, 132
Sustainable agriculture, 168–74. *See also* Agriculture
Symbiosis, 119
Symons, Julian, 20, 22
Syria, 143
Szary, Ron, 57
Szilard, Leo, 105

Taboos, 158, 161
Tadmor, Naphtali, 39
Talmud, xi, 113
Taraxacum officinale Weber (dandelion), 99–103, 191
Taxonomy (taxonomists), 67, 70–72, 116, 122
Teaching, 3, 44–46, 59, 61, 66–72, 88–89, 132, 156, 190, 193. *See also* Education; Universities
Technology, ix–x, 27, 52, 63, 71, 91–94, 107, 109–10, 117, 127, 147, 153, 174, 189–92; agricultual, 167–74; intermediate, 155; low-energy, 120; of destruction, 112–13, 122; space, 79–94. *See also* Revolution, scientific-technical (S-T); Weapons
Teel, D. J., 150
Teilhard de Chardin, Pierre, 187
Television, 14, 29, 33, 89–90
Teller, Edward, 104–13
Terraces, 193
Testament, New, 176
The Natural History of Selborne, 29–33, 123
Thoreau, Henry David, 1, 20
Three Mile Island, 144
Tidestromia oblongifolia, 35
Tigers, Siberian, 181

Toad, common, 15–16, 18, 25, 28
Toronto, 193
Tortuguero, Costa Rica, 3–7
Tramps, 19, 24
Transportation, 174, 192
Trees; of heaven. *See Ailanthus;* planting of, 16–17
Trow, George, 136
Truman, Harry S., 84
Trurl, 187–88
Tsiolkovsky, Konstantin, 81–82, 89–90, 93
Tuberculosis, bovine, 131
Tuitions, 70, 72
Tundra, 141
Turtle; green, 4–7; musk, 41–42, 46; sea, 4–7, 120, 131
2,4-D, 102

Ulam, Stanislaw, 106
Unemployment (unemployed), 189–90
Uniformity, 115. *See also* Diversity, biological; Generality
Universities, 53–56, 63, 66–77, 85, 116–17, 142, 174, 190
University of California; at Berkeley, 105–6; at Davis, 106; at Santa Barbara, 158
University of Chicago, 105
University of Florida, 4
University of Wisconsin at Platteville, 174
Urbanization (urban life), 52, 63, 82, 142, 184, 192
U.S. Steel (USX), 134
Utopia, 24–25, 27, 192

Vallentin, Antonina, 105
Value (of biological diversity), 117–23, 137
Vermin, 186
Verne, Jules, 80
Vicar of Bray, 16
Vietnam (Vietnam War), 88, 102, 106, 111, 142
Vines, 181
Violets, 103

Volkswagen, 192
Von Braun, Werner, 81
Von Ward, Paul, 50
Von Weizsäcker, Carl Friedrich, 104

Walnuts, 17, 167
Walruses, 127
Waples, R. S., 150
War; nuclear, 110–11, 189; Second World, 142, 189. *See also* Vietnam (Vietnam War)
Washington, D.C., 105
Waste, 49, 193
Water hyacinths, 141
Water resources, 165
Waterton, Charles, 32–33
Watson, James D., 123
Watson, Thomas J., 132
Wealth, 49, 114, 157, 191
Weapons; nuclear, 104–7, 110–12, 189; production, 56–57, 190; race (technology), 11, 83–87, 93; systems, 53, 82, 106, 110
Weather forecasting, 148
Webb, James, 85, 87, 93
Weed, 100–102, 186; weeding, 167
Welfare state, 189
Wells, Adrian, 101
Wells, H. G., 27, 80, 90
Westinghouse, 56
West Point, 135
Weyerhauser, 192
Whales, blue, 118
Wheat, 166, 186
White, Gilbert, 29–33, 123
White, Peter, 141
Wiesner, Jerome, 86
Wildcat, 184
Wilderness, 16, 18, 40, 98, 179, 182
Wildlife management, 148, 150–51
Windows, openable, 73–78
Wisconsin, 174
Wolves, 131, 185
Women, 68, 93
Women's movement, 142
Wood, Lowell, 110

Wood Bison National Park, 130
Woodpecker, redheaded, 185
World Bank, 155
World class scholar (wixel), 68–71
Worldwatch Institute, 179
Wren, golden-crowned, 123
Wrightson Island, 128–30

X-ray laser, 110

Yemen, 34
Yertle the Turtle, 42–43
Yosemite Valley, 178

Zooplankton, 149
Zoos, 123, 181–83, 186